Sacred Natural Sites

Guidelines for Protected Area Managers

WWF for a living planet®
Sacred Land Film Project
The Christensen Fund
LTS International
UNESCO United Nations Educational, Scientific and Cultural Organization
MaB Man and the Biosphere Programme
JCU James Cook University
Rainforest CRC
WCPA World Commission on Protected Areas

IUCN (International Union for Conservation of Nature)

Created in 1948, IUCN brings together more than 80 States, 110 government agencies, 800 NGOs, and 10,000 scientists and experts from over 180 countries in a unique worldwide partnership. The Union's mission is to influence, encourage and assist societies throughout the world to conserve the integrity and diversity of nature, and to ensure that any use of natural resources is equitable and ecologically sustainable. The Union is the world's largest environmental knowledge network, and has helped over 75 countries prepare and implement national conservation and biodiversity strategies. The Union is a multicultural, multilingual organization with 1,000 staff located in 62 countries. Its headquarters are in Gland, Switzerland.

Website: www.iucn.org

The World Commission on Protected Areas (WCPA)

The World Commission on Protected Areas (WCPA) is the world's leading network of protected area managers and specialists, with over 1,300 members in 140 countries. WCPA is one of the six voluntary Commissions of IUCN and is administered by the Programme on Protected Areas at IUCN's headquarters in Gland, Switzerland. WCPA's mission is to promote the establishment and effective management of a worldwide representative network of terrestrial and marine protected areas, as an integral contribution to the IUCN mission.

Website: www.iucn.org/themes/wcpa

James Cook University

James Cook University is pleased to be a partner in the production of this internationally important series of guidelines for protected area management. The University offers opportunities for specialised coursework programmes in protected area management at undergraduate and postgraduate levels. JCU has an outstanding international reputation for its research in marine science, marine biology, tropical biology and environmental science as well as for its programmes in associated natural and social sciences. There are many opportunities across a range of disciplines to pursue PhD studies as well as professional masters level programmes (for example the MAppSci in Protected Area Management). JCU has campuses in Townsville and Cairns, adjacent to the Great Barrier Reef and the Tropical Rainforest World Heritage Areas.

Website: www.jcu.edu.au

Contact: Associate Professor Peter Valentine (peter.valentine@jcu.edu.au)

Rainforest CRC

The Cooperative Research Centre for Tropical Rainforest Ecology and Management is a research and education partnership, which brings together a range of experts in an exciting portfolio covering key areas of research including: environmental planning and management in rainforest regions; managing and monitoring impacts arising from rainforest access and visitation; conservation principles and management; evaluating ecosystems goods and services in a dynamic landscape; and Aboriginal capacity building and collaborative management. The Centre's vision is to see the rainforests of Australia managed and utilized in a scientifically sound and sustainable manner to yield economic and social benefits to the community, while ensuring conservation of the unique cultural and natural features of the rainforest.

Website: www.rainforest-crc.jcu.edu.au

Contact: rainforestcrc@jcu.edu.au

Sacred Natural Sites

Guidelines for Protected Area Managers

Task Force on the Cultural and Spiritual Values of Protected Areas
in collaboration with UNESCO's Man and the Biosphere Programme

Robert Wild and Christopher McLeod, Editors
Peter Valentine, Series Editor

World Commission on Protected Areas
Best Practice Protected Area Guidelines Series No. 16

This publication has been made possible in part by funding from The Christensen Fund, LTS International, The Sacred Land Film Project, UNESCO and WWF International.

Published by: IUCN, Gland, Switzerland and UNESCO, Paris, France

Citation: Wild, R. and McLeod, C. (Editors) (2008). *Sacred Natural Sites: Guidelines for Protected Area Managers*. Gland, Switzerland: IUCN.

ISBN–13: 978–2–8317–1039–6

Cover photo: Front: At Devils Tower National Monument in Wyoming, the US National Park Service uses signs to educate tourists and promote respect for Native American spiritual practices © Christopher McLeod.

Back (clockwise from top left): Sunrise behind Agua and Pacaya volcanos, from the summit of Acatenango, in the central highlands of Guatemala, a sacred landscape for the Maya people © Estuardo Secaira; Visitors entering the Dhimurru Indigenous Protected Area (IPA) in Arnhem Land, Northern Territory, Australia, are asked to respect Aboriginal sacred sites and warned of the consequences of entering the area without a permit © Christopher McLeod; In the American Southwest, Hopi elder Dalton Taylor consults with US Forest Service archaeologist Linda Farnsworth at a mining operation that has destroyed a pilgrimage route into the sacred San Francisco Peaks, near Flagstaff, Arizona © John Schaefer; During the Umbanda ceremony in the National Park of Tijuca in Rio de Janeiro, Brazil, offerings are made to *orixás* – the local forest spirits © Marco Terranova.

Layout by: Bookcraft Ltd, Stroud, UK

Produced by: IUCN Publications Services

Printed by: Thanet Press Ltd, Margate, UK

Available from: IUCN (International Union for Conservation and Nature)
Publications Services
Rue Mauverney 28
1196 Gland
Switzerland
Tel +41 22 999 0000
Fax +41 22 999 0010
books@iucn.org
www.iucn.org/publications

The text of this book is printed on 90gsm matt/offset made from low chlorine pulp

Contents

List of figures

List of maps

List of tables

Foreword

One of the most salient forms of culture-based conservation has been the identification and protection of sacred natural sites, which often harbour valuable biodiversity and protect key ecosystems. Indigenous, local and mainstream cultures and spiritual traditions with their respective world views created protected areas long before the advent of the Yellowstone National Park model on which current protected area legislation, policy and practice are mostly based worldwide. Sacred natural sites are indeed the oldest protected areas of the planet.

Yet, unfortunately, many sacred natural sites are at risk. They are subject to a wide range of pressures and threats, external and internal, such as illegal extraction of timber and wildlife, impacts from extractive industries' operations, encroachment by outsiders, disrespectful tourism, poverty and population dynamics, degradation of neighbouring environments, reduction of the availability of lands and resources for traditional peoples and so on.

Sacred natural sites have also been inadvertently integrated in legally declared protected areas by governments, without recognition of the local community values and of the traditional beliefs, practices, skills and knowledge that have sustained the associated locations, cultures and resources. At times, management direction has prevented access to and use of these areas by traditional communities, resulting in violation of indigenous rights, the creation of mistrust and animosity and a lack of local support for the effective management of such sites and areas.

Action is needed for culturally appropriate sacred natural site protection and management. Current conservation agencies and local traditional communities should receive the support they need to work together to face the threats affecting sacred natural sites. Local and indigenous traditional communities should be supported with economic and other resources to assist in their efforts in preserving the environment. Protected area agencies should recognise the cultural and spiritual dimension of sacred natural sites included within their designated boundaries, and recognise the rights and interests of the communities concerned to continue using and managing those sacred sites as places for their cultural and spiritual realization and reverence. National and international organizations should also recognise the skills and knowledge that local and indigenous communities have in managing the resources and areas associated with sacred natural sites. Furthermore, effective action in support of the preservation and effective management of sacred natural sites will have a large impact on enhancing biodiversity conservation, as well as on the long-term vitality of the cultures that have cared for them.

These guidelines are a contribution from the IUCN Task Force on the Cultural and Spiritual Values of Protected Areas and UNESCO's Man and the Biosphere Programme to support the efforts of a wide spectrum of faith groups and indigenous and traditional peoples of the world for the long-term conservation of their sacred natural sites. The guidelines are the result of an international workshop held in Kunming, China in 2003. They were extensively discussed and revised at the World Parks Congress in Durban, South Africa in 2003, and at the International Symposium on "Conserving Cultural and Biological Diversity: The Role of Sacred Natural Sites and Cultural Landscapes", in Tokyo, Japan in 2005. Subsequent to this meeting the guidelines have been restructured and supporting material added to achieve their current format. After four years of field-testing they will be re-evaluated and revised.

Dr Thomas Schaaf
UNESCO Man and the Biosphere (MAB) Programme

"What we're accommodating here at Devils Tower is indigenous peoples' rights to their culture. They're here for the long haul, and they know it. These sacred sites are central to the perpetuation of their culture, and one of our jobs here at Devils Tower is to protect that right."

Deborah Liggett, Superintendent (1994–1997)
Devils Tower National Monument, USA

Acknowledgements

We would like to thank all those many people who contributed to the development of these guidelines.

The idea to produce these guidelines originated with Allen Putney, then Leader of the Task Force on Cultural and Spiritual Values of Protected Areas, who took forward the first draft of the working guidelines, in close collaboration with Thomas Schaaf of the UNESCO Man and the Biosphere Programme. They were joined by Gonzalo Oviedo of the IUCN Secretariat and Christopher (Toby) McLeod of the Sacred Land Film Project. Robert Wild took on the mantle of Task Force leadership and of producing this volume of the IUCN Guidelines in 2007 and acted as lead author and co-editor with Christopher McLeod.

Members of the Task Force, particularly Bas Verschuuren, have played key support roles during the discussions over and development of this articulation of the guidelines, and many have contributed ideas, case study materials and photographs.

We appreciate the many people who have developed and presented more than 50 case studies in various workshops over the past 10 years, and those who have developed the brief case studies presented here.

We would also like to thank the organizers and funders of the workshops that allowed this material to be presented, notably the Ford Foundation. To The Christensen Fund, which made the work on this document possible, we offer our sincere gratitude. Robert Wild and Christopher McLeod are also indebted to our own institutions, LTS International and the Sacred Land Film Project of Earth Island Institute, which have supported us in this endeavour. Finally, we would like to thank the many people who reviewed and commented on the various drafts of the document.

Glossary of working definitions

Many of the terms used in this volume have long been the subject of academic debate that is beyond the scope and objectives of these guidelines. For this reason, we have adopted the working definitions set out below. We are conscious of the fact that there are many, layered meanings to most words, and that many fundamental concepts expressed here do not easily translate from other languages into English (and the reverse is also true). A single language cannot express the riches of some 6,000 languages worldwide. Many cultures have not experienced the separation from nature that is common in the West. Notably, the equivalent word for "nature" in many languages includes the sense of the sacred that it does not in English. Thus, it may be hard to find direct equivalents in many languages for key terms such as "nature", "sacred" and "faith" – let alone other words of recent coinage such as "biodiversity".

Culture. Culture is a set of distinctive spiritual, material, intellectual and emotional features of society or a social group. It encompasses, in addition to art and literature, lifestyles, ways of living together, value systems, traditions and beliefs.[1]

Custodians. Individuals or groups of people, usually within traditional institutions, who have the responsibility to take care of a specific sacred natural site or sites. Custodians may reside either close to or at considerable distance from the sacred natural sites to which they are linked through history, culture, self identification and spiritual practice.

Mainstream faith. The eleven larger faiths which are, in alphabetical order, Bahai, Buddhism, Christianity, Daoism, Hinduism, Islam, Jainism, Judaism, Shinto, Sikhism and Zoroastrianism.

Protected area (IUCN definition). An area of land and/or sea especially dedicated to the protection and maintenance of biological diversity and of natural and associated cultural resources and managed through legal or other effective means (IUCN, 1994).[2]

Protected area (CBD definition). A geographically defined area, which is designated or regulated and managed to achieve specific conservation objectives.

Sacred site. An area of special spiritual significance to peoples and communities.

Sacred natural site. Areas of land or water having special spiritual significance to peoples and communities.

See also 1.2 Use of terms.

Abbreviations

CBD	Convention on Biological Diversity
CCA	Community conserved area
CEPA	Communications, education and public awareness
FPIC	Free, prior and informed consent
IPA	Indigenous protected area
IUCN	International Union for Conservation of Nature
SNS	Sacred natural sites
WWF	World Wide Fund for Nature

1 UNESCO, 2002 http://portal.unesco.org/culture/en/ev.phpURL_ID=13031&URL_DO=DO_TOPIC&URL_SECTION=201.html

2 At the time of going to press a revised IUCN definition is under development but not finalized. The draft is "A clearly defined geographical space, recognised, dedicated and managed to achieve the long-term conservation of nature, associated ecosystem services and cultural values [through legal or other effective means/through state or other effective governance]".

Dedication

We would like to dedicate this volume to the many custodians of sacred natural sites that are so important to environmental and human well-being. May your efforts and resolve to care for and guard these special places continue to gain strength.

PART I

INTRODUCTION AND BACKGROUND

1 Preamble

1.1 Guidelines as a process

These guidelines evolved over the period 2003–2008. Their original focus was the sacred natural sites of indigenous and local communities, and this remains their primary scope. These communities, of which there are many thousands across the globe, usually hold deep sacred values in regard to nature, values that are often focused on and rooted in specific locations. Some indigenous people have such a close relationship to their sacred natural sites that the deterioration or destruction of those sites threatens their very existence. In addition, sacred natural sites related to indigenous and local communities are, in general, more vulnerable and more threatened than sacred natural sites associated with mainstream faiths.

Hundreds of sacred natural sites that are still guarded and maintained by traditional, indigenous custodians have been incorporated into protected areas in recent decades. Consistent with IUCN policies, priority has therefore been given, at this point in guideline development, to the sacred natural sites of indigenous and local communities, which account for an extraordinarily high level of both cultural and biological diversity.

During the process of guideline development, it was recognised that mainstream faiths also care for numerous sacred natural sites and many have profound teachings related to the relationship between humans and nature.[1] We have endeavoured to develop the guidelines in such a way that they are broadly applicable to the sacred natural sites of all faiths. While retaining the focus on local communities and indigenous people, experiences derived from the sacred natural sites of the mainstream faiths have been included in the narrative, as well as a number of case studies. This attempt to incorporate limited experiences of mainstream faiths should be considered preliminary. Further work is needed to analyse and understand the diversity of sacred natural sites revered by mainstream faiths, which comprise the great majority of humankind.[2]

Whether of indigenous, local or mainstream traditions, the management of sacred natural sites in legally protected areas is in its infancy. It is envisaged that experiences in applying these guidelines will be reviewed in four years and the guidelines will then be modified as appropriate.

1.2 Use of terms

A mention is needed regarding the terms used in these guidelines. We have adopted working definitions (see glossary) and accept that these have their limitations. By and large, we have used the Convention on Biological Diversity (CBD) usage of "indigenous and local communities" as defined by Borrini-Feyerabend *et al.* (2004). This does not imply, however, that these definitions are perfect or enjoy complete acceptance. When we say "indigenous" or "local community faith" or "spiritual tradition", we mean faith of an indigenous people under these previous definitions, as expressed through local cultural beliefs and practices. As far as we know, these many faiths or belief systems have not been counted, but they are likely to be in excess of 6,000.[3] In this volume we have adopted the usage of Dudley *et al.* (2005) for, as they put it, "the so-called mainstream faiths". The mainstream faiths, in this context, refer to the eleven faiths that the World Wide Fund for Nature (WWF) and the Alliance for Religion and Conservation (ARC) have been working with over the past two decades and that have produced position statements on faith and ecology (see glossary).

Of course, efforts to put such ideas into categories fail at some point. For example, Shinto and Daoism are undoubtedly indigenous religions of Japan and China, practised by many people indigenous to those countries. But they are not faiths of the more restricted definition of Indigenous People as defined by the United Nations.

1.3 The audience for these guidelines

These guidelines are provided primarily to assist protected area managers, especially those with sacred sites located within the boundaries of their legally established protected areas. Recognising the primacy of traditional custodians in managing their sites, it would be inappropriate for IUCN or UNESCO to provide management advice regarding sacred sites for which custodians have successfully cared for many generations. It is hoped that the guidelines will promote cooperation between protected area managers and custodians of sacred sites towards the enhanced conservation of these special places.

---

1 See Annex 8.

2 The Delos Initiative (http://www.med-ina.org/delos/) of the IUCN Task Force on Cultural and Spiritual Values of Protected Areas, has made significant steps in this regard, particularly in the technologically developed countries, (see the Montserrat (2006) and the Ouranoupolis (2007) statements, Annex 7; Mallarach and Papayannis, 2007; Papayannis and Mallarach, 2008) and the issues will be treated in future Guidelines under preparation by the Delos Initiative.

3 With about 6,800 known languages (Oviedo *et al.*, 2000; Maffi, 2004), and between 5–6,000 Indigenous Peoples (Beltrán, 2000), and many more local communities, this figure is likely to be conservative.

While managers of protected areas are the main focus for the guidelines, it is hoped that they will be of use to a wider group of stakeholders and policy makers. This advice is therefore aimed at:

- managers of individual protected areas with sacred natural sites located either within them or nearby;
- managers of protected area systems who have sacred natural sites within or in the sphere of influence of their network of protected areas;
- natural resource ministries responsible for protected area agencies and systems.

Other stakeholders that may find these guidelines useful are:

- planning authorities responsible for land-use planning outside protected areas;
- traditional custodians who wish to engage with environmental or protected area authorities to increase the protection of their sacred sites, or seek or offer advice about ecological management;
- non-government and other agencies that are providing support to the custodians of sacred natural sites;
- other custodians, governments and biological or cultural support organizations that wish to support the conservation of sacred natural sites.

We have aimed to keep site-based protected area managers clearly in mind in developing the guidelines. In terms of flow, the guidelines develop from the specific to the more general, from the local to the national level. For some of the broader and national level issues, the main role of site managers will be to advocate for policy changes based on the experiences in their individual protected areas.

It should also be noted that we recognise that "technical guidelines" and "best practice" themselves stem from a Western, scientific and reductionist conservation culture,[4] and that attempting to marry these with the holistic and traditional-knowledge based cultures represented at sacred natural sites is ambitious. Increasingly, however, protected area managers are willing to learn from the experience of traditional custodians and, at the same time, custodians of sacred natural sites are requesting advice from ecologists regarding the environmental management of their sites.

1.4 Case studies

With thousands of sacred natural sites worldwide, the selection of examples and case studies for this volume has not been an easy task. Rather than full case studies, what we present here are "snapshots" related to and illustrating, where possible, specific guidelines. These snapshot case studies are derived from the significant (but still relatively small) case study material built up over the last ten years. These case studies were presented at meetings organized by UNESCO, IUCN or collaborating institutions. Specifically, these meetings have been held in India (1998), China and South Africa (2003), Mexico and Japan (2005), Spain (2006), Mongolia, Greece and the UK (2007).[5] Other related work has also been referred to, particularly the important recent work of WWF and the Alliance for Religion and Conservation related to protected areas,[6] as well as their earlier efforts related to faith and nature.[7]

---

[4] MacDonald, 2004.

[5] Ramakrishnan *et al.*, 1998; Harmon and Putney, 2005; UNESCO, 2003 and 2006; Pumarejo and Berges, 2005; Mallarach and Papayannis, 2007; UNESCO, 2007; Papayannis and Mallarach, 2007.

[6] Dudley *et al.,* 2005.

[7] Edwards and Palmer, 1997; Palmer and Finlay, 2003.

2 Introduction

Natural areas that are held to be sacred by peoples are found all across the Earth. This is a widespread phenomenon to be found in almost every country. There are considered to be many thousands of distinct belief systems around the globe and many have ethics related to conservation. Links between faiths and the conservation of land and water were found in every belief system examined by Dudley *et al.* (2005). In these guidelines, those areas have been called **Sacred Natural Sites** and are defined as "areas of land or water having special spiritual significance to peoples and communities".

For many peoples, sacred natural sites are areas where nature, connection to the greater universe, and collective or individual recollections come together in meaningful ways. Sacred natural sites can be the abode of deities, nature spirits and ancestors, or are associated with hermits, prophets, saints and visionary spiritual leaders. They can be feared or they can be benign. They can be areas for ceremony and contemplation, prayer and meditation. For people of no particular faith they often inspire awe and can induce a sense of well-being. They can also hold secular values for history, culture, relaxation and enjoyment. Sacred natural sites can be important places of reference for cultural identity: for an extended family, a clan, a tribe, a religious faith or entire nations that may root their identity in a specific place in nature.

In many societies, traditional sacred natural sites fulfil similar functions as legal protected areas. Due to the spiritual values attributed to these sites, restrictions on access and use often apply, and many such sites remain in a natural or near-natural condition. Here, human disturbance has been reduced or prevented, or careful management has taken place, often for long periods of time, with resulting high levels of biodiversity. At some sacred natural sites, the human influence is greater and these sites may be semi-natural or even heavily modified, but often in ways that retain high levels of biodiversity. Examples of the latter are the globally important ingenious agricultural heritage systems, many of which are attributed spiritual values by the communities that have shaped them.[1] Thus, while many sacred sites have restricted use, others are used on a frequent and on-going basis. Some receive large numbers of visitors which can swell to thousands and, in some cases, millions during pilgrimages.[2]

Clearly, human communities have often not conserved nature, but sacred natural sites represent one example where people generally have done so. Many of these sites have survived environmental degradation because they are deeply embedded in local cultural values and belief systems. They often provide sanctuary to rare or endangered species. For a number of species, their survival is heavily dependent on sacred natural sites. These important places can, therefore, play a significant role in biodiversity conservation by preserving gene pools needed to restore degraded environments.

Natural and semi-natural ecosystems cannot be understood, conserved or managed without recognising how human cultures have influenced and, in some cases, shaped them. Many seemingly untouched ecosystems have actually been the subject of sophisticated management regimes. It is increasingly recognised that biological and cultural diversities are mutually reinforcing and interdependent.[3] Together, therefore, cultural diversity and biological diversity hold a critical key to ensuring resilience in both social and ecological systems – a critical asset in an era of dramatic global change. Sacred natural sites are significant due to their dual character in conserving cultures and protecting nature. Commonly, these sites are rich in biodiversity *because* they are held sacred, not in spite of it. Thus, if nurtured, these special places can contribute meaningfully to both the conservation of biological diversity and the maintenance of cultural identity.

Sacred natural sites can be contained within legally protected areas or they can lie outside legally designated protected area systems. In some cases, protected areas are a small part of a wider sacred landscape. In all of these situations, sacred natural sites pose particular challenges with regard to their recognition, conservation and management. It must be acknowledged that many protected areas have been superimposed over the traditional use areas of local communities, indigenous and traditional peoples, and mainstream faiths. In setting up protected areas around the world, the values and importance of sacred places and traditional uses have often been ignored, thus affecting the fundamental rights of local cultures. This situation, sometimes based on opposing world views, has many times led to conflict and mistrust, and created obstacles to the development of constructive relationships and cooperation between indigenous or traditional peoples, local communities, faith groups and conservation agencies.

Where sacred places occur outside protected areas in the wider countryside, or in rural or urban landscapes, they may be jeopardised by a whole range of pressures ranging from land-use change, poverty, urbanization, demographic movements and

1 Koohafkan and Boerma, 2006.
2 E.g. Motonaka, 2005; Wickramasinghe, 2005.
3 This is well recognised for example for Aboriginal fire management in Australia, (Rose, 2002, and Szabo and Smyth, 2003) and increasingly more widely (Tucker and Grim, 2001).

growth, mining, agriculture and forestry. Additional threats include overuse, inappropriate visitation, looting of burial grounds and archaeological sites, vandalism and neglect.

Protected areas are usually created to protect valued landscapes, wildlife and biological diversity. They are commonly based on Western scientific models that discriminate against local cultures and traditional knowledge. Sometimes people are even relocated to create new parks. Sacred natural sites, therefore, may not be on the agenda of the political and environmental agencies that establish the protected areas and develop management plans. The purpose of these guidelines is to correct those oversights by suggesting management options and procedures that involve local people in protecting sacred natural sites in protected areas.

Increasingly, the cultural and spiritual values of protected areas are being recognised as an important overall function of protected areas. In the Durban Accord, a statement from the 3,000 participants of the Vth IUCN World Parks Congress, protected areas are described as:

> "Those places most inspirational and spiritual, most critical to the survival of species and ecosystems, most crucial in safeguarding food, air and water, most essential in stabilizing climate, most unique in cultural and natural heritage and therefore most deserving of humankind's special care".

Sacred natural sites themselves demonstrate that humankind's special care is not new and the inspiration and spiritual values that certain places evoke were recognised hundreds if not thousands of years prior to the modern conservation movement. The care and protection afforded to many sacred natural sites by indigenous, local communities and mainstream faiths have over the last century been reconfirmed by national protected area authorities, by overlaying the sacred sites with protected area systems. Unfortunately, this process has, in many cases, been at the expense of the traditional custodians and it has disrupted traditional management.

Sacred natural sites remain outstanding assets of the whole of humanity, and due to the threat of on-going degradation and loss, are deserving of urgent support. The purpose of these guidelines, therefore, is to increase that support by providing guidance based on the shared experiences of field practitioners, protected area managers and traditional custodians.

3 Background to Sacred Natural Sites

This section explores in more depth elements of sacred natural sites brought out in the introduction, and sets out the frame of reference for the guidelines themselves (Sections 4 and 5).

3.1 What is a sacred natural site?

For the purposes of these guidelines, sacred natural sites (SNS) are defined as areas of land or water having special spiritual significance to peoples and communities.[1] Many sacred natural sites are areas of great importance for the conservation of biodiversity. In fact, for many communities it is difficult to separate the reasons for protecting the spiritual connections between people and the earth, and for conserving biodiversity in their lands.[2]

Sacred natural sites can be considered a subset of "sacred sites", which may be defined as areas of special spiritual significance to peoples and communities. The wider category of sacred sites may include primarily natural areas (such as forests), but also human-built or monumental areas (such as temples). Many built or monumental sacred sites are themselves located in areas with high natural values that are often recognised by the faith concerned. IUCN focuses on areas that are primarily natural, as they link to its mission, but generally supports the cause of conserving both monumental and natural sacred sites as valuable elements of human cultures. In accordance with their spiritual beliefs, many traditional communities throughout the world have given a special status to natural sites such as mountains, volcanoes, rivers, lakes, springs, caves, forest groves, ponds, coastal waters and entire islands. Many of these have been set aside as sacred places. The reasons for their sacredness are diverse. They may be perceived as abodes of deities and ancestral spirits; as sources of healing water and medicinal plants; places of contact with the spiritual realm, or communication with a "more-than-human" reality; and sites of revelation and transformation. They are sometimes the burial grounds of ancestors, places of pilgrimage, the locale of a temple, shrine or church, or sites associated with special events, saints and spiritual leaders. Over wider landscapes, particular plant and animal species may also be considered as sacred by some communities. While many sacred natural sites have historical significance, they are not static in time or space; new sites can be created in response to changing circumstances and environment.[3]

3.2 Custodians and other stakeholders

A defining feature of sacred natural sites is that people have cared for them, often for a very long time. These are people who have acted as guardians and custodians of the spiritual, cultural, biological and other values of such sites and who are usually closely identified with them. Often these guardians work collectively and have formed a variety of institutions to look after the site. For clarity, the single term "custodians" is used throughout these guidelines. "Custodians", however, covers in a single word, complex and often multi-layered management, ownership, and institutional situations. Very careful analysis is needed to understand the custodial situation of a particular sacred natural site. Many custodians reject the Western concept of "ownership" but embrace community responsibility for taking care of land, water, sacred sites and other "resources".

The custodians of some sacred natural sites can thus be a clearly identifiable group of people with well-defined authority and leadership structures. In other circumstances, custodians can be multiple, dispersed and diffuse. In some cases, the traditional custodians do not live near the sites and can actually live considerable distances away.

The custodial situations of individual sacred natural sites are therefore unique and need to be approached with care. Custodians may range from an extended family, several clans, tribes or other indigenous groups, whole communities, multiple indigenous ethnic groups, churches, temples, monastic orders, groups of monastic orders, sects, and groups from multiple religions.

Identifying and interacting with custodians of sacred natural sites often requires great sensitivity, respect and trust building, sometimes in historically difficult, politically charged and very tense situations. The legitimacy, and sometimes even the authenticity, of individuals or groups to be recognised as custodians cannot be assumed. In the case of indigenous peoples,

---

1 Oviedo and Jeanrenaud, 2006.
2 Oviedo, 2001.
3 Oviedo and Jeanrenaud, 2006.

governments have sometimes imposed upon them forms of leadership that act in parallel to traditional organizations. In dealings with the wider society, indigenous people have faced bribery, corruption and even the creation of fake organizations and leaders.[4] The situation can also be complex with local communities and mainstream faiths. An important challenge for custodian communities of all sorts is to ensure that their systems of decision making are genuinely representative and developed in ways that are inclusive of, and accountable to, members of their communities.[5]

At many sites, multiple custodians have operated harmoniously and respectfully over long periods of time. At other sites, there may be long-standing historical differences and conflict. Conflict is also not new within protected area contexts, and conflicts over sacred natural sites between various parties, including with the state, may have a long and deep history, and often present serious challenges for protected area managers.[6] Protected area agencies working with sacred natural sites clearly need to be able to identify, hire and retain staff with the requisite sensitivity, negotiation and conflict management skills.

Custodians of sacred natural sites are central stakeholders, but there are many others. These include the protected area agencies themselves, owners of the land where the site exists (who are not always the custodians), neighbouring land owners, lease or concession holders, wider community members, government agencies, conservation organizations, visitors, and the national and international public. (See Box 7 for guidance on stakeholder analysis).

In some cases, a site may be sacred to more than one community of different faiths, with different rules of use and management, with accommodations made between these communities. In some cases the community that held a site sacred no longer exists and it is common for the care of such sites to be maintained in some form either by local practice or government agencies. Many ancient sacred sites often retain some spiritual meaning for people – even if they no longer know anything about the original belief systems.

3.3 The origins and cultural legacy of sacred natural sites

Knowledge of the origins of sacred natural sites, and the traditions associated with them, often lies in the stories and myths of communities that hold them sacred. These frequently relate to the genesis of the peoples themselves, for example, the origin myths of several peoples in Asia are linked with mountains that are still held to be sacred in those countries. Sacred natural sites are, in fact, involved in the founding histories of many communities of both indigenous and mainstream faiths. Often there is a direct link between the gods and the people, with the ancestors of particular traditions being the offspring, consorts or messengers of the gods. Some of these origin stories of sacred natural sites have been written down, and there are cases where communities have documented records that go back many generations. Most, however, reside in the oral traditions of the communities that hold them sacred, while for some sacred natural sites their origin has become obscured by time. Increasingly, knowledge of sacred sites is being supplemented by the archaeological record. In terms of age, many sites appear to be in the range 200–400 years old, while quite a number date back thousands of years. Paintings on cliffs in Australia's Kakadu National Park, for example, have been dated at 25–40,000 BCE. In contrast some faiths create new sacred natural sites whenever their adherents settle somewhere new, and so not all sacred sites are old.

The story of sacred natural sites is an element of the complex relationship between humans and nature. At sacred natural sites the relationship is, by definition, reverential. This reverence is frequently for nature itself. Alternatively, it can apply to significant events, historical associations, settings for ritual communion with deities, or for mediation with the cosmos:

> "... whether nomadic hunter-gathers or settled farmers, these varied ecocultural groups, surviving under relatively harsh environmental conditions, have always looked at nature with awe and reverence".[7]

Thus many societies have maintained, at least to some extent, a connection with nature, often mediated through sacred natural sites. Even for those cultures and faiths now quite removed from nature, sacred natural sites have relevance. In secular Western societies, for example, the origins of the concept of "sanctuary", and the widespread custom of throwing coins into urban fountains, can be traced to practices occurring at ancient sacred natural sites.

3.4 Conservation and biodiversity values

There is an expanding body of research demonstrating that:

a) many sacred natural sites support high levels of biodiversity;

4 Colchester and Ferrari, 2007.
5 Colchester and Ferrari, 2007.
6 Taylor and Geffen, 2003.
7 Ramakrishnan *et al.*, 1998, in Ramakrishnan *et al.*, 2003.

b) these biodiversity levels are often much higher than in the areas around them where habitats have been significantly modified by various types of land use; and

c) these biodiversity levels are sometimes higher than those in similar habitat types found nearby.[8]

It is increasingly recognised that this fact is not incidental but is due to the protection afforded to these sacred areas by the custodian communities themselves. In situations where these sites exist in largely modified landscapes, where protected areas are few, they may be the only example resembling the original vegetation, and represent a gap in biodiversity conservation. In what is probably the most comprehensive review of this neglected area, Bhagwate and Rutte (2006) examined 98 references on sacred natural sites. They found examples from all across the globe. Examples of studies that demonstrate the higher biodiversity values for sacred natural sites include sacred groves in India,[9] Kenya,[10] Ghana[11] and Tanzania.[12]

In some cases, specific species of both animals and plants survive exclusively in sacred natural sites. For example, the only remaining Ghanaian population of the true Mona monkey subspecies *Cercopithecus mona mona*, lives in a small (28ha) sacred grove at Tafi Atome, in the Volta region. The Mona monkeys are associated with sacred values and are considered sacred themselves. It is taboo to kill them. A festival to celebrate the monkeys takes place every February, and increasingly the community is benefiting from tourism revenue.[13]

The links between the conservation values of sacred natural sites and the management practices of their custodians are elegantly described by Erjen Khamaganova, an indigenous leader of Buryat people from the Lake Baikal region of Russia:

"More and more people are recognizing the correlation between the spiritual work of a native person in a sacred place and the corresponding higher level of biodiversity associated with such sites. The devout attitude of native persons to sacred places and both their inability and impossibility to destroy the inherent harmony of these places has created conditions for the conservation of biodiversity. Sacred sites are the cornerstone of our cultures' world views and native philosophies, not mere conservation activities. The efforts of indigenous people to protect biodiversity and to preserve our cultures are interconnected and inseparable. Rare species of flora and fauna exist today by virtue of the special place in traditional cultures and their protection and regeneration within sacred sites. In the course of centuries, indigenous people have been protecting sacred sites with special care and thus protecting and promoting sacred birds and animals, sacred plants and trees and associated landforms and waterscapes".[14]

As mentioned, sacred sites are often relatively small, and represent fragments of previously larger areas of vegetation. Research has shown that while biodiversity survives much better in larger areas, small patches of habitat are very important and can harbour a disproportionate number of species. The emphasis on habitat restoration can only be expected to grow in coming years. Some communities have indicated an interest in enlarging the area of their sacred sites, although the socio-economic conditions will need to be favourable to achieve this.[15] Sacred natural sites are likely to become important nodes for habitat restoration, sources of rare species, and links between sites, ecological corridors and support/buffer zones. As such, they can play a critically important role in ecological conservation in and around legally protected areas.

Further research into biodiversity and sacred natural sites is needed to expand the number of local studies and to build these into more comprehensive regional and global summaries.

3.5 Landscape and aesthetic values

Many sacred natural sites are considered "beautiful", and it is likely that their beauty has, in part, led to them being attributed spiritual values or "power". Dramatic landscapes, imposing rock formations, mountains, waterfalls, ancient trees, all these aspects of nature can stir the spirit. The beauty and power of nature is a recurring theme in cultures that value sacred sites and landscapes. A human mark is sometimes placed in these spiritual landscapes. A number of traditions, modest in the extreme, leave no obvious trace. Others may build a small cairn, construct a shrine of local materials, or hang prayer flags. Some cultures are more imposing, constructing larger edifices, such as stupas, temples, mosques and monasteries. Often these are built "organically" over many years, are made of local materials and earth colours, and they fit into the landscape. Despite their larger size they are still dwarfed by the imposing scale of nature surrounding them.

8 E.g. Tiwari *et al.,* 1998.
9 Tiwari *et al.,* 1998.
10 Githitho, 2005.
11 Ormsby, 2007.
12 Mgumia and Oba, 2003.
13 Ormsby, 2007.
14 Khamaganova, 2007.
15 Ormsby, 2007.

3.6 Numbers and extent of sacred natural sites

Sacred natural sites occur at a variety of scales. They can be as small as a single tree or rock formation, or can extend to an entire mountain range. In some cases, whole landscapes are regarded by a community as sacred, containing within them areas of more special sacred focus.

The total number of sacred natural sites is unknown. It is estimated, however, that there are between 150–200,000 sacred groves in India alone.[16] In Ghana, there are an estimated 1,900 sacred groves, while in Mongolia there are an estimated 800 sacred natural sites. A global estimate in excess of a quarter of a million sacred natural sites may not be unrealistic.

In terms of land area it has been estimated that 400–800 million hectares of forest are owned or administered by local communities,[17] and a proportion of this almost certainly will be considered to be sacred lands. Additionally, it has been estimated that the property owned by mainstream faith communities covers 7% of the land area of the planet. Not all of this will be considered sacred, but because it is under the control of faith communities these lands could be more consciously managed for biodiversity conservation.[18] Thus, sacred natural sites can make a significant contribution to the global target of reducing the rate of biodiversity loss.

3.7 Threats to sacred natural sites and rates of loss

As is true for much of the natural world, sacred natural sites are suffering widespread losses. The Millennium Ecosystem Assessment (2005) identifies the broader impact of human use on cultural ecosystem services:

> "Human use of all ecosystem services is growing rapidly. Approximately 60% (15 out of 24) of the ecosystem services evaluated in this assessment (including 70% of

Box 1. A summary of the sources of loss of Indian sacred groves

1. *Commercial forestry:* Many sacred groves were destroyed under commercial forestry operations.

2. *Development projects:* Some of the sacred groves that fell under government vested lands were destroyed when townships grew. Railroads and highways have also taken their toll of many sacred groves. Others were flooded by large dam projects.

3. *Shift in belief system:* In some cases, conversion to other religions has resulted in the degradation of sacred groves.

4. *Sanskritisation:* In many places, local folk deities have been, and continue to be, replaced with Hindu gods and goddesses. This has resulted in the erection of temples in sacred groves.

5. *Pilgrimage and tourism:* The integrity of many groves with regional or pan-Indian character has suffered due to the influx of large numbers of pilgrims and tourists.

6. *Removal of biomass:* In many sacred groves, removal of biomass and cattle grazing is permitted. Continuation of these practices over generations has resulted in the dwindling of the groves.

7. *Encroachment:* Many instances are reported where the groves have been encroached upon by local communities and/or by various government departments, as well as by people migrating from outside.

8. *Modernisation and market forces:* The most recent threat to sacred groves comes from the process of modernisation. Local traditions are being challenged by westernised urban cultures. Modern education systems fail to instil respect for local traditions. As a result, the institution of sacred groves is losing its cultural importance for the younger generation of local people. The spread of market economy has resulted in the denial and erosion of separate identities of local communities. The lure of short-term commercial gains has prompted destruction of the traditional resource base, including the sacred groves.

9. *Fragmentation and perforation:* Many of the sacred groves are fragmented and perforated by roadways, extension of power lines, or reclaimed land for agriculture. Such fragmentation leads to loss of species, and disruption of ecological functions.

Source: Malhotra *et al.*, 2001 in Gokhale, 2003.

[16] Gokhale, 2003; Chatterjee *et al.,* 2004.
[17] Molnar and Scherr, 2003 in Barrow and Pathak, 2005.
[18] See http://www.arcworld.org/projects.asp?projectID=9

Figure 1. The growth of global protected areas over time (from Chape *et al.*, 2003)
Note: 38,427 PAs covering approx. 4 billion km² have no known date of creation and are not included

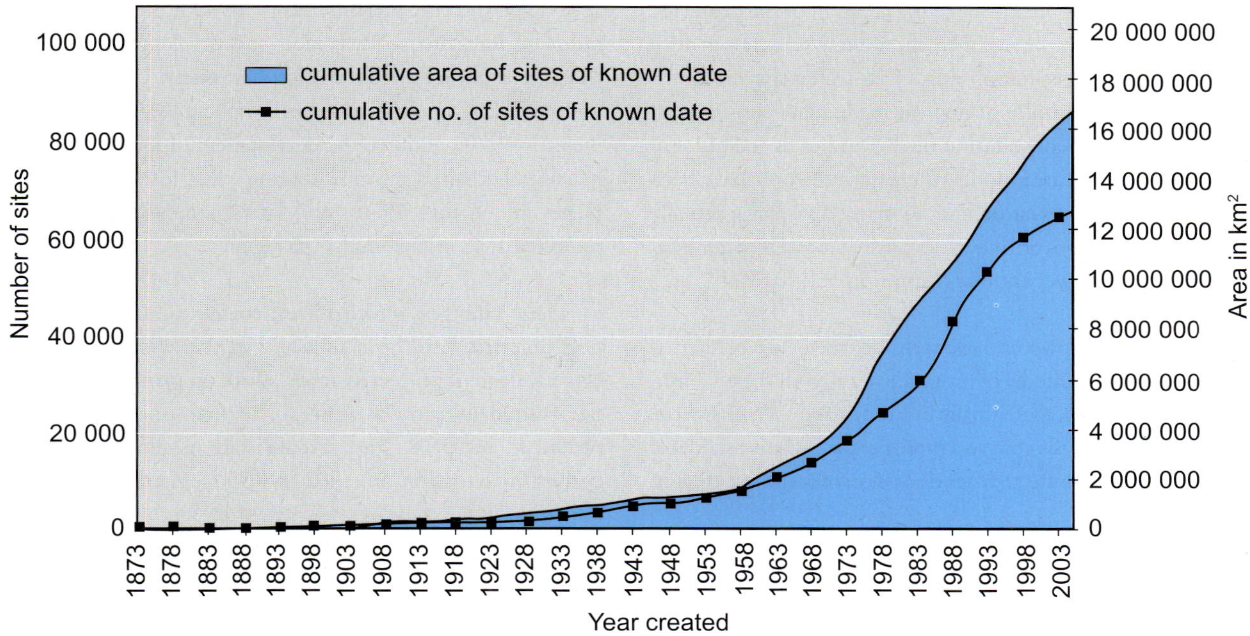

> regulating and cultural services) are being degraded or used unsustainably … Although the use of cultural services has continued to grow, the capability of ecosystems to provide cultural benefits has been significantly diminished in the past century … The rapid loss of culturally valued ecosystems and landscapes can contribute to social disruptions and societal marginalisation".

Malhotra *et al.* (Box 1) have identified a number of threats to India's sacred groves, an analysis that is helpful because most of the threats and causes are common phenomena pertaining to all countries.

There is an urgent need to better understand the extent, patterns and causes of loss of sacred natural sites, to enable action to reduce these losses and so conserve biodiversity and cultures. The Millennium Ecosystem Assessment concluded that many communities place high value on these services and that the impact of their loss is keenly felt:

> "The impact of the loss of cultural services is particularly difficult to measure, but it is especially important for many people. Human cultures, knowledge systems, religions and social interactions have been strongly influenced by ecosystems. A number of the Millennium Ecosystem Assessment sub-global assessments found that spiritual and cultural values of ecosystems were as important as other services for many local communities, both in developing countries (the importance of sacred groves in India, for example) and industrial ones (the importance of urban parks, for instance)".

3.8 Sacred natural sites and the modern protected area movement

The modern protected area movement is estimated to be about 140 years old, beginning with the designation in the USA of Yosemite Valley as public trust land in 1864, followed by the establishment of Yellowstone National Park in 1872.[19] These protected areas were established largely as wilderness areas, entailing the exclusion of native people in what is now referred to as the "Yellowstone Model".[20] Compared to sacred natural sites that were established thousands of years ago, modern protected areas are newcomers onto the scene of land management. As the World Wide Fund for Nature (WWF) concluded in their 2005 report, *Beyond Belief*: "Sacred sites are probably the oldest method of habitat protection on the planet".

During the rapid expansion and development of modern protected areas (see Figure 1), there has been a myopia – or deliberate disregarding – of pre-existing community use of the lands

[19] IUCN, 1994; Stevens, 1997.
[20] Stevens, 1997.

designated for "protection". For example, in only one of the eleven case studies in the IUCN Best Practice Guidelines on Indigenous and Traditional Peoples and Protected Areas,[21] had the communities been formally consulted prior to the creation of the protected area. Although this sample size is small, it likely reflects a general pattern of minimal consultation during protected area establishment. Many other examples of poor or inadequate consultation exist. The low level of consultation further indicates the hitherto exclusionary protected area model where native people have been seen as inimical to conservation and, in many cases, have actually been removed from government-established protected areas at considerable social and economic disruption and hardship.[22]

The last 50 years, however, have seen the emergence of alternatives to the exclusionist model of protected areas, with the establishment of protected landscapes initially in Europe and then elsewhere, and more recently with the shift to a more people-inclusive model of protected areas.[23] The protected area discussion over the last 20 or so years has hinged largely on this shift and many important studies have documented this evolution.[24] The movement to be more inclusive started with community education programmes in the 1980s, and then expanded in the 1990s with the widespread application of integrated conservation and development programmes. These generally followed the "biosphere reserve concept" developed under UNESCO's Man and the Biosphere (MAB) Programme in the early 1970s, and also the development of joint or collaborative management arrangements, starting with forestry in India and Nepal in the mid 1970s, and later more widely in parks and protected areas in the South Asia region.

These changes, which have become mainstream conservation practice, have been reflected in the evolution of IUCN's classification of protected areas. Work on protected area categories started as early as 1969. The first full articulation was produced in 1978, and it consisted of ten categories. The current articulation emerged in 1994, when six categories of

Table 1. The classification of management category and governance type (from Borrini-Feyerabend *et al.*, 2004)

Governance Types	A Government Managed Protected Areas			B Co-Managed Protected Areas			C Private Protected Areas			D Community Conserved Areas	
Protected Area Categories	Federal or national ministry or agency in charge	Local/municipal ministry or agency in charge.	Government-delegated management (e.g. to an NGO)	Transboundary management	Collaborative management (various forms of pluralist influence)	Joint management (pluralist management board)	Declared and run by an individual landowner	… by non-profit organizations (e.g. NGOs, universities, cooperatives)	… by for profit organizations (e.g. individual or corporate landowners)	Declared and run by indigenous peoples	Declared and run by local communities
Ia – Strict Nature Reserve											
Ib – Wilderness Area											
II – National Park											
III – Natural Monument											
IV – Habitat/Species Management											
V – Protected Landscape/Seascape											
VI – Managed Resource Protected Area											

21 Beltrán, 2000, p.39.
22 West and Brechin, 1991; Adams and McShane, 1992; Stevens, 1997.
23 Phillips 2003, in Borrini-Feyerabend *et al.*, 2004.
24 West and Brechin, 1991; Wells and Brandon, 1992; Kempf, 1993; Stevens, 1997.

Table 2. Examples of sacred natural sites in IUCN categories (from Verschuuren *et al.*, 2007)

Ia	Strict Nature Reserve: protected area managed mainly for science		
	Sri Lanka	Yala National Park	Significant to Buddhists and Hindus and requiring high levels of protection for faith reasons.
	Russian Federation	Yuganskiy Kanthy	Significant to Christianity. The protected area has been created around Lake Numto – a Khanty and Nenets sacred place – in Beloyarsk region.
Ib	Wilderness Area: protected area managed mainly for wilderness protection		
	Mongolia	Bogd Khan Mountain	The mountain is significant to Buddhism and previously to shamanism. The mountain has been officially designated as a sacred mountain by the state. Evidence exists of a wilderness area declaration dating from 1294.
	Mongolia	Dornod Mongol	Significant to Buddhism. Vangiin Tsagaan Uul (White Mountain of Vangi) is a sacred Buddhist peak within the reserve.
II	National Park: protected area managed mainly for ecosystem protection and recreation		
	Malawi	Nyika National Park	Large area containing four sacred sites, which local people still use for rainmaking ceremonies.
	Japan	Kii Mountains National Parks and WHS	Several Shinto and Buddhist temples, sacred sites and pilgrimage trails for both faiths in continuous use for over one millennium.
	India	Great Himalayan National Park	Includes many places of religious importance for Hinduism.
III	Natural Monument: protected area managed mainly for conservation of specific natural features		
	Cambodia	Phnom Prich Wildlife Sanctuary	A small area within the sanctuary is a sacred forest and therefore a natural monument (another example are the kaya forests of Kenya).
	Russian Federation	Golden Mountains of Altai WHS	Sacred to indigenous Altaians and many different faiths including Buddhist, Christian and Islamic.
	Greece	Mount Athos WHS	Stronghold of Orthodox Christianity including 15 monasteries and a number of hermitages with over one millennium of continuous monastic activity.
	Spain	Montserrat Nature Reserve and Natural Park	Christian monastery with centuries-old hermitages which has been a pilgrimage centre since the 14th century. Today it is the most heavily visited protected area in Spain.
IV	Habitat/Species Management Area: protected area managed mainly for conservation through management intervention		
	Lebanon	Qadisha Valley and the Forests of the Cedars of God WHS	Sacred forest to the Christian Maronite Church, including a significant monastery, hermitages, and residence of religious authorities.
	Borneo	*tembawang* gardens	Some sacred sites will need continual intervention or even to be planted – such as the tembawang gardens that contain high levels of biodiversity.
	Sri Lanka	Peak Wilderness Park (Sri Pada-Adams Peak)	Sacred natural site for Islam, Buddhism, Hinduism and Christianity, attracting many pilgrims of all these faiths.

Table 2. Examples of sacred natural sites in IUCN categories (cont.)

V	Protected Landscape/Seascape: protected area managed mainly for landscape/seascape conservation and recreation		
	China	Xishuangbanna National Park	Landscape with several sacred sites (groves and mountains), which have long been managed by the community and are part of an important and biologically rich cultural landscape.
	Romania	Vanatori Neamt Natural Park	The spiritual heart of Romania, including 16 Christian monasteries, along with outstanding wildlife: European bison, brown bear and wolf populations.
VI	Managed Resource Protected Area: protected area managed mainly for the sustainable use of natural ecosystems		
	Ecuador	Cayapas Mataje	Sustainable use area said to contain the world's tallest mangroves and known for important spirit dwellers that are worshipped by local people.
	USA	San Francisco Peaks, National Forest	Sacred to over one dozen Native American tribes.
	Egypt	St Catherine Area WHS, Mt Sinai	Mount Sinai is sacred to Judaism, Christianity and Islam. The ancient monastery of St Catherine is a World Heritage Site.

protected area were established.[25] The six categories generally represent a gradation of increasing human disturbance in natural ecosystems. It was at this time that "protected landscapes" and "managed resource protected areas" achieved equal standing with the first four more strictly protected categories.[26]

Overarching these six categories is the IUCN protected area definition:

> " … areas of land and/or sea especially dedicated to the protection and maintenance of biological diversity, and of natural and associated cultural resources and management through legal or other effective means" (IUCN, 1994 and 2008).[27]

The IUCN definition (both existing and revised versions) explicitly recognises that in addition to biodiversity and nature, cultural resources and cultural values are worthy of protection. Implicit in this language is the recognition that cultural management methods other than legal recognition can also be effective protective mechanisms (for example, customary management).

Thus, the awareness that communities have been managing areas of land for conservation purposes over long periods of time has been a lesson that the protected area movement has had to re-learn. This was first formally recognised as recently as 2003 at the Durban World Parks Congress. Here, for the first time, the concept of Community Conserved Area (CCA) gained widespread acceptance from the protected area community as a newly recognised form of governance. Table 1 overlays the six IUCN protected area categories with the typology of CCA governance stemming from that meeting.[28]

This new governance typology has been instrumental in formalizing international recognition of the important historical role that communities have played in land management. The elucidation and application of the CCA concept at national and local levels is in progress, and its implications for protected area management are evolving. Specifically, the relationship of sacred natural sites and community conserved areas is in the process of illumination.

Recent work has shown that sacred natural sites occur in all six IUCN management categories (see Table 2). In their report, *Beyond Belief*, Dudley *et al.* (2005) presented information on 100 protected areas that contain sacred natural sites. This review identified the IUCN category under which the protected areas were designated (74% of sites recorded this information, Figure 2) and also recorded the international designation of the site under one of the international protected area conventions or programmes. The 100 sites selected for study were chosen to illustrate particular features relevant to the subject matter and were not a randomly selected sub-sample of protected areas

25 IUCN, 1994.
26 Brown *et al.*, 2005.
27 At the time of going to press a revised IUCN definition is under development but not finalized. The draft is "A clearly defined geographical space, recognised, dedicated and managed to achieve the long-term conservation of nature, associated ecosystem services and cultural values [through legal or other effective means/through state or other effective governance]".
28 Borrini Feyerabend, 2004.

Figure 2. Distribution of 74 protected areas containing sacred natural sites, by IUCN categories (from Dudley *et al.*, 2005)

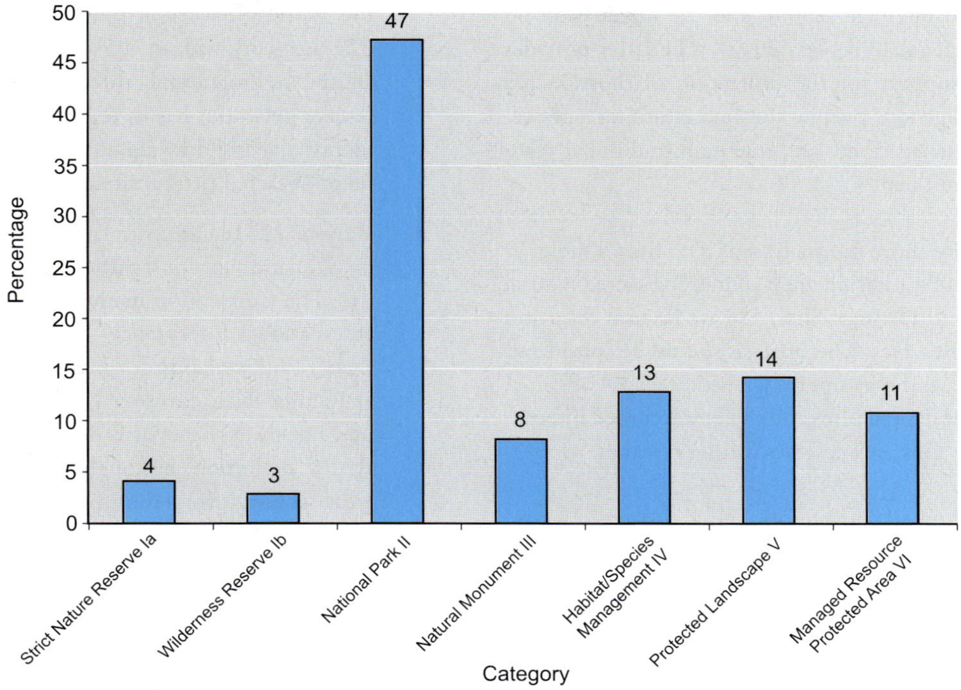

containing sacred natural sites. This analysis can therefore only be seen as indicative rather than representative of the pattern of sacred natural sites and IUCN categories, which is worthy of further research and study.

A significant number of sacred natural sites were located within the four most strictly protected IUCN categories I–IV, that is: strict nature reserve, wilderness area, national park, or species/habitat management area. This would indicate that *modern protection is reinforcing a pre-existing cultural mechanism to minimize human access through prohibitions related to sacred natural sites.* A key function of traditional management systems has often been to restrict access to areas that are regarded to have sacred values.

The conclusion of this improved (although still far from complete) understanding of the pattern of sacred sites and IUCN protected area categories is that while the IUCN categories *may reflect a reduction of human disturbance of ecosystems, this is not necessarily a reflection of an absence of earlier human intervention or management in these areas.* Cultural and traditional attributes, mediated specifically through sacred natural sites, have in some cases been responsible for the reduced human disturbance. Cultural and traditional attributes are, therefore, not only applicable but highly relevant to IUCN Categories I–IV.

Cultural and spiritual values were not well reflected in the 1994 version of the IUCN protected area guidance for these four categories, and this guidance is now under revision. It is now recognised that all protected areas should aim, where appropriate, to: "conserve natural and scenic areas of national and international significance for spiritual, cultural and scientific purposes".[29] We recommend explicit and stronger recognition and supportive management of cultural and spiritual values and sacred natural sites within all IUCN categories.

Greater recognition that sacred natural sites can occur in all IUCN categories allows flexibility in the ways that protected area categories can be used to support the conservation of sacred natural sites. Quite a number have been recognised as national parks and natural monuments (categories II and III), while explicit use of category Ia – strict nature reserve, is being discussed for sacred natural sites in Madagascar (Dudley, *pers. com.*) and is likely to have wider application. Within protected areas where higher levels of access are allowed, zoning could be used to reinforce the restricted access needed to conserve the cultural values of some sacred natural sites.

3.9 International recognition of sacred sites

At the international level, protected areas are supported by a number of programmes, conventions and declarations. These are, in the order of their establishment, the Man and the Biosphere Programme (1970), the Convention on Wetlands, also known as the Ramsar Convention (1971), the World

[29] IUCN, 2008.

Heritage Convention (1972), the Convention on Biological Diversity (1992), the Convention for the Safeguarding of Intangible Cultural Heritage (2003), and the Declaration on the Rights of Indigenous Peoples (2007). The latter provides reference to and support for the protection of those sacred natural sites of indigenous peoples through affording a greater level of rights to manage them and maintain traditional practices associated with them.

Considering once more the analysis of data from Dudley *et al.* (2005), of the 100 protected areas containing sacred natural sites, 17% are World Heritage Sites, 5% are Ramsar Sites and 3% are Biosphere Reserves. This analysis should be considered as very preliminary, and may not fully reflect the level of sacred natural sites in these designations and thereby accorded international recognition. This area deserves further research.

These programmes, conventions and declarations are described in more detail below:

- **MAB 1970:** The Man and the Biosphere Programme established the biosphere reserve model, which set the standard for integrating human needs into protected area management. The establishment of biosphere reserves that are recognised under UNESCO's Man and the Biosphere Programme, are highly innovative and demonstrate new approaches to conservation and sustainable development. They are under national sovereign jurisdiction, yet share their experience and ideas nationally, regionally and internationally within the World Network of Biosphere Reserves. There are 529 sites worldwide in 105 countries. The Man and the Biosphere Programme, through meetings of biosphere reserve networks, has in recent years taken the lead in identifying and drawing out lessons from the management of sacred natural sites.
- **Ramsar 1971:** The central messages of the Convention on Wetlands are conservation, and wise or sustainable use. The Convention maintains a List of Wetlands of International Importance – currently there are more than 1,708 wetlands designated for special protection as Ramsar Sites, covering 153 million hectares. These may not always necessarily be protected areas under the IUCN definition, although they often overlap with areas covered by other protective designations. A number of sacred natural sites fall within Ramsar-recognised wetlands of international importance. While sacred values do not have thus far a significant profile under the convention, improved guidance on cultural values is currently under development.
- **World Heritage 1972:** Through the Convention, UNESCO seeks to encourage the identification, protection and preservation of cultural and natural heritage around the world that is considered to be of outstanding value to humanity. The Convention has 184 state

Box 2. Key elements of the United Nations Declaration on the Rights of Indigenous Peoples of relevance to sacred natural sites

"Recognising that respect for indigenous knowledge, cultures and traditional practices contributes to sustainable and equitable development and proper management of the environment ...

Article 11

1. Indigenous peoples have the right to practise and revitalize their cultural traditions and customs. This includes the right to maintain, protect and develop the past, present and future manifestations of their cultures, such as archaeological and historical sites, artefacts, designs, ceremonies, technologies and visual and performing arts and literature.
2. States shall provide redress through effective mechanisms, which may include restitution, developed in conjunction with indigenous peoples, with respect to their cultural, intellectual, religious and spiritual property taken without their free, prior and informed consent or in violation of their laws, traditions and customs.

Article 12

1. Indigenous peoples have the right to manifest, practise, develop and teach their spiritual and religious traditions, customs and ceremonies; the right to maintain, protect, and have access in privacy to their religious and cultural sites; the right to the use and control of their ceremonial objects; and the right to the repatriation of their human remains ...

Article 25

Indigenous people have the right to maintain and strengthen their distinctive spiritual relationship with their traditionally owned or otherwise occupied and used land, territories, waters and coastal seas and other resources and to uphold their responsibilities to future generations in this regard.

Box 2. Key elements of the United Nations Declaration on the Rights of Indigenous Peoples of relevance to sacred natural sites (cont.)

Article 26

1. Indigenous peoples have the right to the lands, territories and resources which they have traditionally owned, occupied or otherwise used or acquired.

2. Indigenous peoples have the right to own, use, develop and control the lands, territories and resources that they possess by reason of traditional ownership or other traditional occupation or use, as well as those which they have otherwise acquired.

3. States shall give legal recognition and protection to these lands, territories and resources. Such recognition shall be conducted with due respect to the customs, traditions and land tenure system of the indigenous peoples concerned.

Article 29

1. Indigenous people have the right to the conservation and protection of the environment and the productive capacity of their lands or territories and resources. States shall establish and implement assistance programmes for indigenous people for such conservation and protection, without discrimination …

Article 32

1. Indigenous peoples have the right to determine and develop priorities and strategies for the development or use of their lands or territories and other resources.

2. States shall consult and cooperate in good faith with the indigenous peoples concerned through their own representative institutions in order to obtain their free and informed consent prior to the approval of any project affecting lands or territories and other resources, particularly in connection with the development, utilization or exploitation of mineral, water or other resources.

3. States shall provide effective mechanisms for just and fair redress for any such activities, and appropriate measures shall be taken to mitigate adverse environmental, economic, social, cultural or spiritual impact.

Article 34

Indigenous peoples have the right to promote, develop and maintain their institutional structures and their distinctive customs, spirituality, traditions, procedures, practices and, in the cases where they exist, juridical systems or customs, in accordance with international human rights standards."

signatories and provides for the designation and listing of World Heritage Sites. It is incumbent on the international community as a whole to protect World Heritage Sites irrespective of the territory on which they are located. Currently listed are 660 cultural sites, 166 natural sites, with 25 being of mixed cultural/natural designation. Many World Heritage Sites contain sacred natural sites and landscapes. While focusing on tangible heritage, the World Heritage Convention has increasingly recognised intangible values.[30]

- **CBD 1992:** Signed by 150 government leaders at the 1992 Rio Earth Summit, the Convention on Biological Diversity is dedicated to promoting sustainable development. It recognises that biological diversity involves more than plants, animals and micro-organisms and their ecosystems – it is also about people, the need for food security, medicines, fresh air and water, shelter, and a clean and healthy environment. The convention has emphasised traditional knowledge and protected areas. Although it does not have a specific programme devoted to sacred natural sites, CBD has developed the *Akwé: Kon Voluntary Guidelines for the Conduct of Cultural, Environmental and Social Impact Assessment regarding Developments Proposed to Take Place on, or which are Likely to Impact on, Sacred Sites and on Lands and Waters Traditionally Occupied or Used by Indigenous or Local Communities.*[31]

30 Rössler, 2003.
31 Secretariat of the Convention on Biological Diversity, 2004.

- **Living Heritage 2003:** The 2003 Convention for the Safeguarding of Intangible Cultural Heritage (ICH) or living heritage, addresses domains that include: oral traditions; the performing arts (such as traditional music, dance and theatre); social practices, rituals and festive events; knowledge and practices concerning nature and the universe; and traditional craftsmanship.[32] It is, therefore, concerned with many practices that are related to sacred natural sites. The domain of "knowledge and practices concerning nature" is of particular relevance. Prior to the establishment of the convention, UNESCO ran a programme proclaiming "Masterpieces of the Oral and Intangible Heritage of Humanity" in which masterpieces related to sacred natural sites were recognised.
- **Declaration on the Rights of Indigenous Peoples:** This United Nations Declaration, endorsed in 2007 (Annex 3), provides a framework for the rights of Indigenous Peoples and includes reference to:

  > "recognising that respect for indigenous knowledge, cultures and traditional practices contributes to sustainable and equitable development and proper management of the environment".

Box 2 reproduces the key articles and statements related to sacred natural sites.

At the global level, therefore, there is a broad framework for considering cultural and spiritual values and recognising and preserving sacred natural sites within protected areas. Now that the Living Heritage convention has come into force, and the Declaration on the Rights of Indigenous Peoples has been endorsed, there is the enhanced opportunity for these mechanisms to work together in a coordinated way toward the support of sacred natural sites and their custodians.

[32] The convention came into force in April 2006 and 87 nations have signed it as of December 2007.

PART II

GUIDELINES AND CASE STUDIES

Part two of these best practice guidelines are made up of:

Section 4: the guidelines themselves consisting of six principles and 44 guidelines; and

Section 5: elaboration on the guidelines in a narrative text along with relevant case studies.

These guidelines are aimed at enhancing the recognition and sympathetic management of sacred natural sites that are located in legally designated protected areas, as well as advocating for productive and respectful collaboration between protected area managers and traditional custodians. In the best cases, sacred natural sites are supported by government protected area systems and managers; however, more often, sacred natural sites are overlooked and in some cases custodians and protected area managers are in conflict.

The primary audience for these guidelines is the managers of individual protected areas and managers and planners of protected area systems. It would be inappropriate for IUCN or UNESCO to directly advise traditional custodians on the management of sacred sites for which they have successfully cared for many generations. It is hoped, however, that custodians from indigenous and traditional communities and mainstream faiths, along with their support organizations, will find the guidelines useful to improve interaction with protected area managers and to better integrate ecological concerns into sacred site management.

It is hoped that other stakeholders, such as natural resource ministries, state planning departments and private protected area managers will also find these guidelines useful, and that they will play a supporting role to custodians who are protecting and managing sacred natural sites located within unprotected landscapes.

If the guidelines are successful in their intent it is anticipated that lasting and productive partnerships will ensure sacred natural sites and their managing cultures survive long into the future.

[**Top**] At Mount Shasta in northern California (USA), the late Winnemem Wintu leader Florence Jones opens a four-day fire ceremony in the Shasta-Trinity National Forest. Logging, cattle grazing and off-road vehicle harassment have been problems at the traditional ceremonial site. The Winnemem opposed construction of a ski resort on their sacred Mount Shasta in the 1990s and the U.S. Forest Service denied the permit for the project, in part due to the spiritual concerns of Native Americans. *(Photograph by Christopher McLeod)*

[**Above**] In South America, a ceremony in the Andes honours the *Apu* (deity) embodied in the sacred mountain. Ritual gatherings often serve as community meetings to discuss management of some of the planet's original, oldest protected areas (see Case study 16). *(Photograph by Oscar Minera ©UNEP / Topham / The Image Works)*

[Top] "House of Spirits" – Mijikenda elders at Chizia Cha Nyere, a sacred site where ceremonies start in the Kaya forests, Kenya. *(Photograph ©WWF-Canon / Elizabeth Obel-Lawson)*

[Right] Mijikenda tour guide, at left, and visitors at Kaya Kinondo, Kwale District. One of the rules established by Kaya elders is that visitors must wear a black traditional *kikoi* when entering the sacred forest as a sign of respect. Visitors must also be accompanied by an authorized guide. Other rules include a rest day for the forest every four days, a tourism zone, and limits to where photographs can be taken (see Case study 5). *(Photograph by Robert Wild)*

[Below, left] In India, many villagers have begun to proudly display signs to their sacred groves in Kodagu, Karnataka state, following the sacred groves festival in 2000 (see Case study 7). *(Photograph by Shonil Bhagwat)*

[Below, right] The annual ritual of Kovilkadu worships deities in sacred groves (known as *kovilkadu* in the Tamil language) in Tamil Nadu state in southern India. *(Photograph courtesy of the National Museum of Mankind, IGRMS, Bhopal, India)*

[Top left and top right] At Rainbow Bridge National Monument (USA) and Waipi'o Valley, Hawai`i (USA), protected area managers educate visitors and safeguard sacred natural sites using a variety of signage. *(Photographs by Christopher McLeod)*

[Above and left] The rock feature in the photo above at right is part of a site known as *Gaynada*, which is "giant trevally dreaming" or *Nguykal*, sung by the Gumatj, Mangalili and Wanguri clans of northeast Arnhem Land, Australia. It is an example of a sacred site under pressure from the impact of a nearby mining town as evidenced by the vehicle tracks running past it. Camping and swimming are prohibited near the rock outcrop. The area is managed with a permit system enforced by the Dhimurru and Parks and Wildlife Service of the Northern Territory Rangers (see Case study 15). *(Photographs by Jane Dermer, Dhimurru Land Management Aboriginal Corporation, above, and Christopher McLeod, left)*

[Top] In Grand Canyon National Park (USA), Hopi elder Dalton Taylor explains to National Park Service archaeologist Jan Balsom that tourists are removing eagle feather prayer offerings from Hopi shrines in the canyon. As a result of this consultation, the Park Service relocated a hiking trail away from the shrine and the disturbances stopped. Film footage from the consultation was used in a television documentary to educate the public about sacred site issues. *(Photograph by Christopher McLeod)*

[Right] Guillermo Rodriguez-Navarro, UNESCO's Thomas Schaaf and Rogelio Mejia Izquierdo of the Confederación Indígena Tayrona discuss expanding the boundaries of Colombia's Sierra Nevada de Santa Marta Biosphere Reserve and National Park to include more of the sacred sites of the Tayrona people (see Case study 10). *(Photograph by Christopher McLeod)*

[Bottom right] A sacred waterfall in Montagne d'Ambre National Park, Madagascar, is introduced to visitors by the park's warden, at right. Local people believe the spirits of their ancestors dwell in trees, water and rocks. According to oral tradition, these spirits once lived in the surrounding area known as *Antsiranana*. As the forests were cut, the ancestral spirits sought refuge at Amber Mountain. Today, residents of the region come to the falls to ask for blessings of vitality, fertility and purification. *(Photograph by Nigel Dudley)*

[**Above**] At left is the painting of the goddess Pele exhibited in Hawai`i Volcanoes National Park's Visitor Center prior to 2005. At right, the new image selected from dozens of paintings submitted by local artists to represent a more native view of the creator of the land. *(Paintings by David Howard Hitchcock, left, and Arthur Johnsen, right)*

[**Below**] Tourists and photographers visiting Uluṟu-Kata Tjuṯa National Park in Australia are discouraged from viewing sensitive sites and are not allowed to take photographs in certain areas (see Case study 12). *(Photograph of sign taken in 1991 by Christopher McLeod, used with permission of Parks Australia, Uluṟu-Kata Tjuṯa National Park)*

[**Top**] There are approximately 1,900 sacred groves in Ghana, characterized by remnant mature forest patches containing rare biodiversity, often close to villages and within heavily utilized landscapes. Pictured here are the sacred forest of the Boabeng-Fiema Monkey Sanctuary, and the true Mona monkey from Tafi Atome Monkey Sanctuary. This is the only Ghanaian population of this subspecies of the true Mona monkey, which are themselves considered sacred. *(Photographs by Alison Ormsby, right, and Michael Scace, left)*

[**Above**] As a strategy to enhance protection, shaman Maria Amanchina and cultural expert Maya Erlenbaeva map sacred sites that occur outside formal protected area boundaries in the Altai Republic, southern Siberia, Russia. *(Photograph by Christopher McLeod)*

[**Left**] Prehistoric rock carvings, or petroglyphs, are emblems of Altaian cultural heritage, but they are frequently stolen and sold on the black market. The horns of this mountain sheep were damaged when a would-be thief tried to remove the image. The Chui Oozy Nature Park, Altai Republic, Russia, was established by local elders to protect such petroglyphs. *(Photograph by Christopher McLeod)*

[**Above and left**] Mongolia's Bogd Khan Mountain is associated with the life of Chinggis (Ghengis) Khan and has been a nationally protected sacred natural site since 1778. It is now part of the extensive Khan Khentii Mountain Protected Area. After many years of suppression, ceremonies have been revived, led by local Buddhist lamas. The ceremonies honour the deities of the mountain and petition against drought and heavy snow. The group that performs the ritual at the most sacred area at the top of the mountain returns, led by monks. Third person from left is Mr. J. Boldbaatar, Director, Khan Khentii Special Protected Area, and on his right is the first modern-day park ranger (see Case study 13). *(Photographs by Robert Wild)*

[**Left**] The Rila Monastery in Bulgaria was founded in the 10th century by St. Ivan of Rila and is located at the heart of the magnificent Rila Mountains, home to brown bear, wolf and boar. Rila Monastery Natural Park aims to guarantee and preserve the unity between nature and the monastery (see Case study 14). *(Photograph by Nenko Lazarov, from www.imagesfrom bulgaria.com)*

[**Left**] The unique rock formations of Meteora, Thessaly, Greece, attracted hermits and ascetics in the 11th century and monasteries were built in the 14th century. The dramatic, towering rocks are surrounded by oak forests and are home to threatened plants and animals such as the Egyptian Vulture and European Wolf. Meteora is a World Heritage Site listed for both cultural and natural values. *(Photograph by Thymio Papayannis)*

[**Top left**] A masked Gule Wamkulu dancer of the Nyau brotherhood of the Chewa people in Malawi emerges from a sacred grove to participate in a public event (see Case study 9). *(Photograph by Karen Edwards)*

[**Top right**] A ceremony honouring sacred Lake Baikal in Russia is conducted by Altan-Erdeni, spiritual leader of nine Buryat clans. *(Photograph by Vladimir Chenkirov)*

[**Above left**] Offerings of flowers are left in the waters of sacred Chichabal lagoon, a crater lake in the western highlands of Guatemala (see Annex 1). During two days in the spring, 5,000 people come to the shores of the lagoon to perform Maya-Mam ceremonies and pray for plentiful rains and abundant crops. *(Photograph by Estuardo Secaira)*

[**Above right**] On Mt. Ausangate, south of Cusco, Peru, considered one of the holiest mountains of the Andes, a holy man makes an offering to the *Pachamama*, or Mother Earth. *(Photograph by Allen Putney)*

[**Bottom, right**] Young Australian Aboriginal people continue their traditional responsibilities to care for cultural landscapes and sacred sites by learning the songs and dances that connect them to their country. *(Photograph by Christopher McLeod)*

4 Principles and Guidelines for the Management of Sacred Natural Sites Located in Legally Recognised Protected Areas

A working version of the guidelines (7 pages) was published in 2006[1] and has also been available on the internet since then. The present version builds upon the initial concepts and incorporates feedback received regarding the working version. It is envisaged that the current guidelines will be tested in field situations and will be reviewed and revised over the next four years.

In their current form, the guidelines are relatively detailed and prescriptive. The 44 guidance points are grouped into six principles. In terms of flow, they generally develop from the specific and local to the more general and national level. In relation to some of the guidelines at the regional or national scale, it is recommended that individual protected area managers advocate for appropriate, relevant policy changes that will improve management of sacred natural sites locally, nationally and globally.

4.1 Principles

Principle 1	Recognise sacred natural sites already located in protected areas.
Principle 2	Integrate sacred natural sites located in protected areas into planning processes and management programmes.
Principle 3	Promote stakeholder consent, participation, inclusion and collaboration.
Principle 4	Encourage improved knowledge and understanding of sacred natural sites.
Principle 5	Protect sacred natural sites while providing appropriate management access and use.
Principle 6	Respect the rights of sacred natural site custodians within an appropriate framework of national policy.

4.2 Guidelines

Principle 1 Recognise sacred natural sites already located in protected areas.

Guideline 1.1 **Natural and cultural values:** Recognise that sacred natural sites are of vital importance to the safeguarding of natural and cultural values for current and future generations.

Guideline 1.2 **Ecosystem services and human well-being:** Recognise that sacred natural sites have great significance for the spiritual well-being of many people and that cultural and spiritual inspiration are part of the ecosystem services that nature provides.

---

[1] UNESCO, 2006, pp. 326–331.

Guideline 1.3 **Recognition:** Initiate policies that formally recognise the existence of sacred natural sites within or near government or private protected areas and affirm the rights of traditional custodians to access and play an appropriate, ideally key, role in managing sacred natural sites now located within formal protected areas.

Guideline 1.4 **Consultation:** Include the appropriate traditional cultural custodians, practitioners and leaders in all discussions and seek their consent regarding the recognition and management of sacred natural sites within or near protected areas.

Guideline 1.5 **Holistic models:** Recognise that sacred natural sites integrate social, cultural, environmental and economic values into holistic management models that are part of the tangible and intangible heritage of humankind.

Principle 2 **Integrate sacred natural sites located in protected areas into planning processes and management programmes.**

Guideline 2.1 **Park planning:** Initiate planning processes to revise management plans to include the management of sacred natural sites located inside protected area boundaries.

Guideline 2.2 **Identify sacred natural sites:** Where secrecy is not an issue and in close collaboration and respecting the rights of traditional custodians, identify the location, nature, use and governance arrangements of sacred sites within and around protected areas as part of a participatory management planning process.

Guideline 2.3 **Respect confidentiality:** Ensure that pressure is not exerted on custodians to reveal the location or other information about sacred natural sites and, whenever requested, establish mechanisms to safeguard confidential information shared with protected area agencies.

Guideline 2.4 **Demarcate or conceal:** Where appropriate and to enhance protection, either clearly demarcate specific sacred natural sites, or alternatively, to respect the need for secrecy, locate sacred natural sites within larger strictly protected zones so exact locations remain confidential.

Guideline 2.5 **Zoning:** Establish support, buffer and transition zones around and near sacred sites, especially those that are vulnerable to adverse external impacts.

Guideline 2.6 **Linkages and restoration:** Create ecological corridors between sacred natural sites and other suitable areas of similar ecology for connectivity, and in degraded landscapes consider restoring sacred natural sites as an important initial step to reviving a wider area.

Guideline 2.7 **Ecosystem approach**: Adopt the ecosystem approach as the key strategy for the integrated management of land, water and living resources that promotes conservation and sustainable use in an equitable way and also includes cultural and spiritual values.

Guideline 2.8 **Landscape approach**: Take a landscape approach to sacred natural sites, recognising their role in wider cultural landscapes, protected area systems, ecological corridors and other land uses.

Guideline 2.9 **Support development planning recognition:** Development planning authorities are the main planners of land use in areas outside many protected area systems. Seek their and other stakeholders' support for the recognition of sacred natural sites in the wider countryside.

Guideline 2.10 **Protected area categories and governance**: Recognise that sacred natural sites exist in all of the IUCN protected area categories and governance types, and that those that fall outside formal protected area systems can be recognised and supported through different legal and traditional mechanisms according to the desires of their custodians, including as community conserved areas when appropriate.

Guideline 2.11 **International dimension:** Recognise that some sacred natural sites, and the cultures that hold them sacred, cross international boundaries and that some may be within or may surround existing or potential transboundary peace parks.

Principle 3 **Promote stakeholder consent, participation, inclusion and collaboration.**

Guideline 3.1 **Prior consent**: Ascertain the free, prior and informed consent of appropriate custodians before including sacred natural sites within new formal protected areas and protected area systems and when developing management policies affecting sacred places.

Guideline 3.2 **Voluntary participation:** Ensure that state or other stakeholder involvement in the management of sacred natural sites is with the consent and voluntary participation of appropriate custodians.

Guideline 3.3 **Inclusion:** Make all efforts to ensure the full inclusion of all relevant custodians and key stakeholders, including marginalized parties, in decision making about sacred natural sites, and carefully define the processes for such decision making, including those related to higher level and national level policies.

Guideline 3.4 **Legitimacy**: Recognise that different individuals and groups have different levels of legitimacy and authority in decision making about sacred natural sites.

Guideline 3.5 **Conflict management:** Where relevant and appropriate, use conflict management, mediation and resolution methods to promote mutual understanding between traditional custodians and more recent occupants, resource users and managers.

Principle 4 **Encourage improved knowledge and understanding of sacred natural sites.**

Guideline 4.1 **Multidisciplinary approach:** Promote a multidisciplinary and integrated approach to the management of sacred natural sites calling on, for example, local elders, religious and spiritual leaders, local communities, protected area managers, natural and social scientists, artists, non-governmental organizations, and the private sector.

Guideline 4.2 **Integrated research:** Develop an integrated biological and social research programme that studies biodiversity values, assesses the contribution of sacred natural sites to biodiversity conservation, and understands the social dimension, especially how culturally rooted behaviour has conserved biodiversity.

Guideline 4.3 **Traditional knowledge:** Consistent with article 8(j) of the Convention on Biological Diversity (CBD), support the respect, preservation, maintenance and use of the traditional knowledge, innovations and practices of indigenous and local communities specifically regarding sacred natural sites.

Guideline 4.4 **Networking:** Facilitate the meeting of, and sharing of information between, traditional custodians of sacred natural sites, their supporters, protected area managers and more recent occupants and users.

Guideline 4.5 **Communication and public awareness:** Develop supportive communication, education and public awareness programmes and accommodate and integrate different ways of knowing, expression and appreciation in the development of policies and educational materials regarding the protection and management of sacred natural sites.

Guideline 4.6 **Inventories:** Subject to the free, prior and informed consent of custodians, especially of vulnerable sites and consistent with the need for secrecy in specific cases, carry out regional, national and international inventories of sacred natural sites and support the inclusion of relevant information in the UN World Database on Protected Areas. Develop mechanisms for safeguarding information intended for limited distribution.

Guideline 4.7 **Cultural renewal:** Recognise the role of sacred natural sites in maintaining and revitalizing the tangible and intangible heritage of local cultures, their diverse cultural expressions and the environmental ethics of indigenous, local and mainstream spiritual traditions.

Guideline 4.8 **Intercultural dialogue:** Promote intercultural dialogue through the medium of sacred natural sites in efforts to build mutual understanding, respect, tolerance, reconciliation and peace.

Principle 5 **Protect sacred natural sites while providing appropriate management access and use.**

Guideline 5.1 **Access and use:** Develop appropriate policies and practices that respect traditional custodian access and use, where sacred natural sites fall within formal protected areas.

Guideline 5.2 **Visitor pressures:** Understand and manage visitor pressures and develop appropriate policies, rules, codes of conduct, facilities and practices for visitor access to sacred sites, making special provisions for pressures brought about by pilgrimages and other seasonal variations in usage.

Guideline 5.3 **Dialogue and respect:** Encourage ongoing dialogue among the relevant spiritual traditions, community leaders and recreational users to control inappropriate use of sacred natural sites through both protected area regulations and public education programmes that promote respect for diverse cultural values.

Guideline 5.4 **Tourism:** Well managed, responsible tourism provides the potential for economic benefits to indigenous and local communities, but tourism activities must be culturally appropriate, respectful and guided by the value systems of custodian communities. Wherever possible, support tourism enterprises that are owned and operated by indigenous and local communities, provided they have a proven record of environmental and cultural sensitivity.

Guideline 5.5 **Decision-making control:** Strong efforts should be made to ensure that custodians of sacred natural sites retain decision-making control over tourist and other activities within such sites, and that checks and balances are instituted to reduce damaging economic and other pressures from protected area programmes.

Guideline 5.6 **Cultural use:** While ensuring that use is sustainable, do not impose unnecessary controls on the careful harvest or use of culturally significant animals and plants from within sacred natural sites. Base decisions on joint resources assessments and consensus decision making.

Guideline 5.7 **Protection:** Enhance the protection of sacred natural sites by identifying, researching, managing and mitigating overuse, sources of pollution, natural disasters, and the effects of climate change and other socially derived threats, such as vandalism and theft. Develop disaster management plans for unpredictable natural and human caused events.

Guideline 5.8 Desecrations and re-sanctifying: Safeguard against the unintended or deliberate desecration of sacred natural sites and promote the recovery, regeneration and re-sanctifying of damaged sites where appropriate.

Guideline 5.9 Development pressures: Apply integrated environmental and social impact assessment procedures for developments affecting sacred natural sites and in the case of the land of indigenous and local communities support the application of the Convention on Biological Diversity's Akwé: Kon Guidelines for minimizing the impacts of development actions.

Guideline 5.10 Financing: Where appropriate, pay due attention to the suitable financing of sacred natural site management and protection, and develop mechanisms for generating and sharing revenue that take into account considerations of transparency, ethics, equity and sustainability. Recognise that in many parts of the world poverty is a cause of the degradation of sacred natural sites.

Principle 6 Respect the rights of sacred natural site custodians within an appropriate framework of national policy.

Guideline 6.1 Institutional analysis: Understand traditional management institutions and enable and strengthen the continued management of sacred natural sites by these institutions. Make appropriate arrangements for the adoption and management of sacred natural sites that have no current custodians, for example by heritage agencies.

Guideline 6.2 Legal protection: Advocate for legal, policy and management changes that reduce human and natural threats to sacred natural sites, especially those not protected within national protected areas and other land planning frameworks.

Guideline 6.3 Rights-based approach: Root the management of sacred natural sites in a rights-based approach respecting basic human rights, rights to freedom of religion and worship, and to self-development, self-government and self-determination as appropriate.

Guideline 6.4 Confirm custodians' rights: Support the recognition, within the overall national protected area framework, of the rights of custodians to their autonomous control and management of their sacred sites and guard against the imposition of conflicting dominant values.

Guideline 6.5 Tenure: Where sacred natural sites have been incorporated within government or private protected areas in ways that have affected the tenure rights of their custodians, explore options for the devolution of such rights and for their long-term tenure security.

5 Principles, Guidelines, Discussion and Case Studies

This section discusses and explains the principles and guidelines and illustrates them with case studies. The section starts with a box that provides a checklist of actions that a manager of a protected area containing a sacred natural site (SNS) or pilgrimage route might follow. These are cross referenced to specific guidelines.

Box 3. Checklist for managers of protected areas (PA) containing sacred natural sites or pilgrim routes.

Action	Guideline cross reference
1 Assess for the presence of sacred sites in the protected area. Find out if any sacred natural sites exist in the protected area. These may be well known or protected area staff may not be aware that they exist. It is not always necessary or appropriate for park managers to know exactly where the sites are located, but to know that they exist.	1.1, 1.2, 1.3, 2.2
2 Identify and engage with the main custodians. Ascertain who are the legitimate custodians, bearing in mind that custodians might live at some distance from the protected area. Engagement with custodians can be difficult especially if there is a history of conflict. Ensure prior consent for engagement and recognise that trust may need to be developed. Establish where possible a participatory process whereby custodians can negotiate with park management with a view to delegated or collaborative management as appropriate.	1.3, 1.4, 3.1, 3.2, 3.3, 3.4
3 Assess the level of conflict, if any. In the case of conflict, assess its level and explore options for conflict management. If conflict is serious seek specialist advice and mediation. Conflict may be caused by the protected area management themselves, in which case assess options for improved management policies, staff training, facilitation etc.	3.5, 4.8, 5.3, 6.3
4 Determine the need for confidentiality. In cases where sites need to be kept confidential, discuss with custodians mechanisms that might best achieve that, including not seeking to find out where the site is actually located but establishing negotiation and feedback mechanisms to determine if management is appropriate.	2.3, 2.4. 3.1, 3.2
5 Assess the level of management any sacred sites may require. Sacred natural sites occur in very diverse situations, with very different management needs. An initial step will be to understand the level of management, the pressures that the SNS might be under and the kinds of engagement that could be required. Impact assessment techniques may be required.	2.2, 4.1, 4.2, 4.3, 5.1, 5.2, 5.3, 5.4, 5.7, 5.8, 5.9, Annex 1
6 Understand the belief systems or faiths involved, recognising the legitimacy of other world views. If the PA managers are not familiar with the belief system or faith related to the site, efforts should be made to understand it and acknowledge its legitimacy. Seek, subject to prior consent, to understand the traditional knowledge of the custodians and the environmental ethics represented by the SNS.	1.1, 1.5, 3.1, 3.3, 3.4, 4.1, 4.2, 4.3
7 Carry out stakeholder analysis. Carry out a stakeholder analysis, formally if appropriate, aiming to understand the key interested parties of the site. As convener of the consultation and negotiation process, analyse the PA management's position and role, and assess the need for involving neutral third parties.	1.4, 3.1, 3.4, 3.5, Box 7

Box 3. Checklist for managers of protected areas (PA) containing sacred natural sites or pilgrim routes. (cont.)

Action	Guideline cross reference
8 Initiate meetings and mediation. Where appropriate initiate a series of meetings aiming to bring together different stakeholders, individually at first but later in larger groupings, using mediation if necessary.	2.7, 3.2, 3.3, 1.4
9 Assess the pattern of access and use. Issues of access and use will be very important for management. These need to be understood preferably via participatory processes. Tourism use can be supportive or can undermine the sacred values. The need for use by the community or faith group may vary at different times of the year.	4.2, 4.3, 5.1, 5.2, 5.4, 5.5, 5.6
10 Incorporate sacred natural sites or pilgrimage routes into PA management plans. Seek to include the empathetic management of the SNS into the management planning process of the protected area. Seek to use participation and adaptive management principles. Consider the side-effects of pilgrimage and other forms of visitation and work with communities and faith groups to minimize detrimental impacts on biodiversity.	2.1, 2.5, 2.7, 2.8
11 Develop protective strategies. Consult with custodians to protect SNS from visitor pressure, vandalism, resource extraction, etc. using zoning, law enforcement, public education, new regulations and laws.	1.4, 2.4, 2.5, 4.5, 5.2, 5.4, 5.7, 5.8, 5.9, 6.2, Annex 1
12 Consider participatory mapping of use. Consider the use of participatory mapping as a way of recognising the use areas of different cultural traditions within the protected area.	2.2, 4.2, 4.3
13 Develop communication, education and public awareness. Determine appropriate channels of communications with stakeholders regarding the regular management of the SNS. Where appropriate, design and implement education and public-awareness activities, for different audiences including the general public.	1.4, 4.5, 4.8, 5.3, 5.4
14 Institute sustainable financing. Seek mechanisms that help finance the management and the care of the SNS or pilgrimage route.	5.10
15 Advocate changes to policies and laws which better manage the SNS and recognise the rights of custodians. Where national policy is not supportive of the effective integration of the SNS within the protected areas, advocate for appropriate changes.	6.1, 6.2, 6.3, 6.4, 6.5

1 Recognise sacred natural sites already located in protected areas.

1.1 Natural and cultural values

Recognise that sacred natural sites are of vital importance to the safeguarding of natural and cultural values for current and future generations.

Sacred natural sites have great importance for both nature and culture. The natural values of sacred natural sites relate to the diversity of life forms, habitats and ecosystems that they support and the landscapes and geology of which they are a part. Sacred places also inspire values that derive from their aesthetic appreciation. In terms of culture, they form part of the belief systems of many indigenous, local and mainstream spiritual traditions; they may be linked to the national identity of many peoples; and they are often used for important ceremonial and religious purposes.

In recent decades, efforts to conserve biodiversity have necessarily taken centre stage in conservation efforts and have become the fundamental purpose of many protected areas. At the same time, there has been a growing appreciation of the contribution that these areas make to national and local economies. For most people, however, protected areas represent a wider set of values, including recreational, spiritual, cultural, identity, artistic, aesthetic, educational, peace, and therapeutic (Harmon and Putney, 2003). The presence of natural sites in protected areas that are sacred to many spiritual traditions and very large numbers of people, adds considerable depth to this broader set of values that protected areas embody.

Case study 1. Tongariro National Park – 110 years of Maori Stewardship in New Zealand

Snowcapped Tongariro and his fellow volcanoes, Ruapehu and Ngauruhoe, rise majestically from the central volcanic plateau of New Zealand's North Island. To the Maori tribes who have inhabited this land since at least the 14th century, Tongariro is *tapu*, sacred. The mountain is regarded as a key part of their history, revered as both a divine ancestor and the home of human ancestors, and it is venerated in legends.

Eastern Polynesians, travelling thousands of miles in large double-hulled canoes, began settling the North Island of New Zealand by about 1300. One of the first canoes was called Te Arawa, and the Ngati Tuwharetoa, one of the *iwi* (tribes) of the area around Tongariro, holds ancestral identification with Ngatoroirangi, high priest and navigator of that canoe. Legend links Ngatoroirangi with the creation of volcanoes and the naming of Tongariro. According to one version, the priest travelled inland to explore the island and claim land for his people. As he climbed to the summit of Tongariro, a strong south wind brought extremely cold weather. Nearly chilled to death and exhausted by the climb, Ngatoroirangi called out for help from his sisters in the far-away Maori homeland, Hawaiki. They came to him in the form of fire under the earth, leaving a trail of geysers and volcanoes in their path and emerging at Tongariro to warm the priest. Thus, the volcanic landscape represents a genealogical link with the historic homeland of Hawaiki, and the mountains are revered as tribal ancestors. The name Tongariro – *tonga* (south wind) and *riro* (seized) – commemorates the cold wind that almost killed Ngatoroirangi.

In 1887, Chief Te Heuheu Tukino IV, facing the encroachment of European settlers, who he felt threatened Maori sacred land, in an astute and insightful move gave the peaks of Tongariro, Ruapehu and Ngauruhoe to the Crown and the people of New Zealand, on the condition that it be protected under the then new concept of "national park". He feared privatization and land ownership would destroy the cultural link to the mountain and displace his people (Te Heuheu, 2005). The Tongariro National Park Act was passed into law in October 1894, making Tongariro the first national park in New Zealand, the fourth in the world, and the first ever to be gifted by a country's indigenous people.

A century later, in 1990, Tongariro National Park was inscribed on the UNESCO World Heritage List solely on the basis of its natural and environmental values without regard for its vital sacred significance to the Maori. With urging from the Maori people, the park was, however, re-nominated for its value as an "associative cultural landscape". Associative cultural landscapes are a new category adopted by the World Heritage Committee in 1992 to recognise sites that possess "powerful religious, artistic or cultural associations of the natural element rather than material cultural evidence, which may be insignificant or even absent". In 1993, Tongariro once more made history, becoming the first site inscribed on the World Heritage register as an associative cultural landscape and thus almost exactly one hundred years after becoming a national park the sacred values to the Maori people were finally recognised internationally.

While Tongariro National Park serves as a model of how environmental conservation and indigenous culture and beliefs can be mutually supportive, the park faces considerable pressure. The late Sir Hepi Te Heuheu, in the centennial year of his grandfather's gift of the sacred peaks to create the national park, said, "The matter of *tapu* is important. We want to see people enjoy the mountain, but we do not want it desecrated".

Tongariro National Park receives approximately one million visitors each year. Skiing, hiking, biking, camping and mountaineering are all popular. Support facilities inside the park include roads, car parks, trails, ski lifts, viewing platforms, trekking huts and a park village that provides lodging and amenities. The primary problems are garbage, erosion on hiking trails, invasive plants, vehicle pollution and an overburdened and outdated sewage system that releases low-quality waste water into streams.

The New Zealand Department of Conservation manages the park and the Maori *iwi* of the Tongariro region, the Ngati Rangi, Ngati Tuwharetoa and Ngati Tahu, are consulted on all significant management issues within the park, especially where cultural values are involved. There are several Maori members on the Conservation Board. The issue of commercialization in the park is of particular concern to Maori tribes. The current Park Management Plan prohibits extension of the amenity areas inside the park and restricts ski fields to 3% of the total area of the park. Issues of commercial guiding are being discussed.

With so much tradition, history and culture tied to the mountain, the Maori want to see these values, as well as the park's physical integrity, preserved for future generations. Since the designation of Tongariro National Park as a cultural landscape, awareness and understanding of Maori cultural values have increased. Maori have been involved in the redevelopment and creation of new displays at the visitor centres that explain the cultural and natural significance of the park and help foster respect for its careful management and conservation. They have also taken part in planning World Heritage celebrations, developing education resources and biodiversity programmes, and assessing concession applications.

Case study 1. Tongariro National Park – 110 years of Maori Stewardship in New Zealand (cont.)

A good example of the collaboration between park management and the Maori *iwi* is the handling of a safety threat after a series of volcanic eruptions in 1995 and 1996 emptied Mount Ruapehu's Crater Lake and accumulated ash deposits at the lake's outlet. Because of the threat of a volcanic mudflow, or lahar, the Department of Conservation undertook an environmental and cultural assessment, in consultation with the *iwi*, of the options to minimize risks to public safety. One option, which entailed bulldozing a trench into the summit of the mountain, was opposed by the Maori *iwi*, who felt it would "challenge the indigenous integrity and strength of the cultural World Heritage status". The conservation minister decided against such work and instead opted to install a state-of-the-art alarm and warning system and to construct a protective bank along the Whangaehu River to prevent a lahar from overflowing onto the highway – a decision that received commendation from the World Heritage Committee for its ethical and cultural sensitivity.

In discussing the important role of indigenous stewardship, Chief Tumu Te Heuheu, son of the late Sir Te Heuheu, said the following:

"Ko Tongariro te maunga
Ko Taupo de moana
Ko Ngäti Tuwharetoa
Ko Te Heuheu te tangata.
Tongariro is our ancestral mountain
Taupo is our inland sea
Tuwharetoa is our tribe
Te Heuheu is the man.

What I intend to do is underline a fundamental principle in the conservation of cultural and biological diversity: the principle of guardianship or stewardship, in the Maori language the principle of *kaitiakitanga.* When it comes to the management of our cultural landscape, all of the above makes it sound easy, but that is not necessarily so. We have found ourselves in accord with the government, who are the managers, most of the time. We have found ourselves in accord with the users of the park most of the time. But on occasions we have to act vigorously in our role of *kaitiakitanga* – as guardians. ... When the government published a draft management strategy embracing the mountain, *Tuwharetoa* did not agree with some of the proposals. *Tuwharetoa* used the process of the courts to ensure we fully exercised our responsibility of *kaitiakitanga.* After much discussion and goodwill we both eventually agreed on a management strategy.

In exercising the full parameters of *kaitiakitanga,* or guardianship, we are also acutely conscious of the need to balance world-wide interest in the landscape with the tribal interests and the maintenance of our cultural veracity. The challenge is not so much to recognise the relationship between the national park and the tribe, for that is now affirmed by World Heritage provisions. The challenge is to embrace and manage global enthusiasm with tribal integrity, so that both can be understood, appreciated and shared by those who will follow us in the years ahead" (Te Heuheu, 2005).

Source: Adapted from Te Heuheu, 2005 and Polidor, Sacred Land Film Project;
http://www.sacredland.org/world_sites_pages/Tongariro.html

As the world's fourth National Park established in New Zealand in 1894, Tongariro represents the long-standing efforts of the Maori people to safeguard the landscape's cultural and spiritual values and to have these recognised in the face of European colonization.

1.2 Ecosystem services and human well-being

Recognise that sacred natural sites have great significance for the spiritual well-being of many people and that cultural and spiritual inspiration are part of the ecosystem services that nature provides.

Many people derive a sense of well-being from being close to nature. This applies to people of many spiritual traditions as well as secular culture. Contact with nature and a sense of place may be a basic need for human well-being and the urban parks movement in the USA has demonstrated the value of planted green space. It is even reported that a picture of a beautiful landscape can enhance well-being. Certainly, visiting national parks and other protected areas offers visitors an opportunity to make a deeper connection with nature that many refer to as "spiritual" or "healing".

For indigenous and local communities certain natural areas have provided the focus for many of their spiritual traditions and are recognised as sacred. The special nature of these places often demands that there be little or no human impact. It often also requires silence and serenity with the beauty of primordial nature and majestic land forms. Similar sentiments inspire practitioners of mainstream faiths. Thus, undisturbed nature and dramatic landscapes, occasionally adorned by few, but beautiful buildings, provide a sense of harmony with nature.

Figure 3. Linkages among biodiversity, ecosystem services and human well-being (from Millennium Ecosystem Assessment, 2005)

This figure depicts the strength of linkages between categories of ecosystem services and components of human well-being that are commonly encountered, and includes indications of the extent to which it is possible for socio-economic factors to mediate the linkage. (For example, if it is possible to purchase a substitute for a degraded ecosystem service, then there is a high potential for mediation). The strength of the linkages and the potential for mediation differ in different ecosystems and regions. In addition to the influence of the ecosystem services on human well-being depicted here, other factors – including other environmental factors as well as economic, social, technological and cultural factors – influence human well-being, and ecosystems are in turn affected by changes in human well-being.

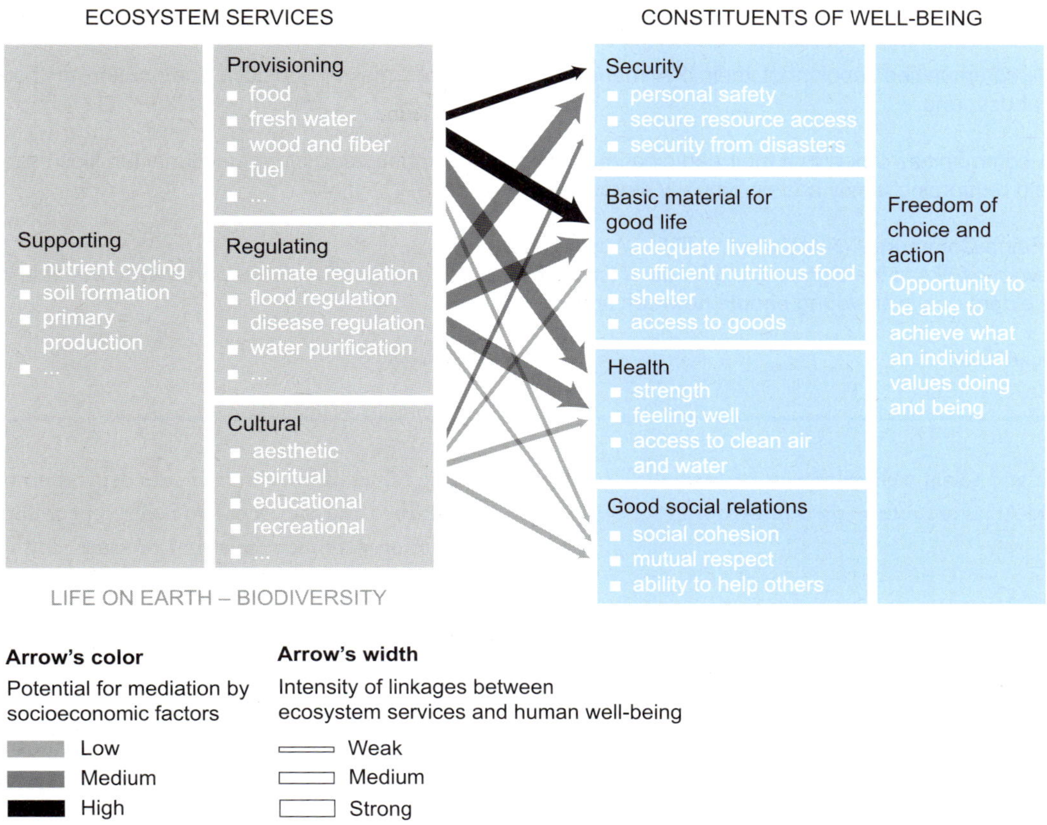

Sacred natural sites are not exclusively valued for their "intangible" services. They are often sources of water – for example, mountain catchments and holy wells, rivers and lakes – and they can also provide medicines, food, ritual paraphernalia and other resources.

Modern economies have consistently undervalued the environmental services that natural systems provide, and this has led to serious deterioration of ecosystems.[1]

The relatively recent emergence of the concept of ecosystem services has started to redress this undervaluation. These services are most often thought of as providing material benefits such as water catchment and coastal protection. The recognition that nature might provide health and well-being has only recently been considered as part of ecosystem valuation. Figure 3 shows that cultural services are now seen as an important component of human well-being. Aesthetic, spiritual, educational and recreational activities in nature cannot easily be substituted by other socio-economic factors. For many communities, sacred natural sites are themselves a focus of that sense of well-being. In the words of Erjen Khamaganova of the Buryat peoples, Russian Federation:

> "Every indigenous nation has created its own system of promoting healthy ways of life … Sacred sites in these systems have always played a particularly important role. When entering a sacred place, a person who is properly educated in the indigenous way does not break any taboos and thus finds him/herself under the patronage of the spiritual owners of that place. A person constantly under pressure from the everyday routine, upon entering the sacred place, finally gets an opportunity to be simply him/herself. The feeling of finding refuge and protection has a tremendously favourable effect on the human mind, body and soul."[2]

1 Millennium Ecosystem Assessment, 2005.
2 Khamaganova, 2007.

Box 4. Positive social impacts of Australian Indigenous Protected Areas (IPA)

The following statistics, generated from internal reporting, support the argument that Indigenous Protected Areas deliver improved social outcomes:

95% of IPA communities report economic participation and development benefits from involvement with the Programme;

60% of IPA communities report positive outcomes for early childhood development from their IPA activities;

85% of IPA communities report that IPA activities improve early school engagement;

74% of IPA communities report that their IPA management activities make a positive contribution to the reduction of substance abuse; and

74% of IPA communities report that their participation in IPA work contributes to more functional families by restoring relationships and reinforcing family and community structures.

The Nepabunna Community (Nantawarrina IPA – Australia's first IPA):
"The Nantawarrina IPA is the biggest thing for us. It turned everything around. The IPA is particularly valuable in helping the community elders deal with young people between school and work".

Source: Gilligan, 2006.

The health and social welfare aspects of Australia's Indigenous Protected Areas are summarised in Box 4.

1.3 Recognition

Initiate policies that formally recognise the existence of sacred natural sites within or near government or private protected areas and affirm the rights of traditional custodians to access and play an appropriate, ideally key, role in managing sacred natural sites now located within formal protected areas.

Sacred natural sites should be officially recognised in culturally appropriate and sensitive ways that enhance the protection of these sites and respect and affirm the rights of their traditional custodians.

Sacred natural sites within government protected areas: Where sacred natural sites occur within established and legally protected areas, their recognition by government authorities, subject to appropriate agreements, can legitimize and formalize the contribution of the traditional custodians to park management and can increase the overall protection of the entire area. This can be realized through engagement and support of a broader community, and so benefit the protected area as a whole. Official recognition may enhance local pride in a sacred site. It may also help to safeguard against the unintended desecration of sacred sites by visitors who are not aware they are in a place considered sacred by indigenous, local or mainstream spiritual traditions. Formal acknowledgement of sacred sites may also help avoid or reduce friction among local communities, conservation agencies and government institutions over land-use practices.

The values of many sacred sites depend on them remaining undisturbed and their location being kept confidential. Official recognition may increase visitor pressure, and thereby increase vulnerability to accidental or intended damage. The decision to recognise a particular sacred natural site should, therefore, be considered very carefully and will only be fully effective if the custodians of the site agree, give prior consent (see guideline 3.1), and express their willingness to support such recognition. Appropriate processes are required to identify the recognised and authentic custodians responsible for a specific site, which, in some cases, can be a complex undertaking (guideline 3.4). It is important, to the extent possible, to recognise and endorse the traditional rules of custodians regarding the management of sacred natural sites (guidelines 2.2 and 6.4). Sometimes, locations must be kept secret (guideline 2.4).

Sacred sites and the establishment of new protected areas: Sacred natural sites may be the focus around which a new protected area is proposed, or sacred natural sites may be located in areas that are proposed for protection based on biodiversity, scenic landscape or other values. In either situation, a full understanding of the nature of the sacred sites, and the engagement, consent and support of their traditional custodians are necessary.

A significant number of cases have shown that the high biodiversity of areas proposed for protected area status derives from the long-standing protection given to such areas by indigenous and local communities. The fact that these communities often do not *mark* their sacred areas with shrines, temples or other outward symbols means that in some situations government and scientific agencies find it difficult to understand and even accept the role that indigenous and local communities have played in conserving biodiversity and other values.

The carefully developed official recognition of sacred natural sites through protected area mechanisms may help to increase the protection of the sacred natural site, while also strengthening overall protection of biodiversity and the wider environment. It will also help to conserve and promote local cultural values and legitimize the role of traditional custodians in protected area management.

Such official recognition will be particularly useful in cases where the natural environment is subject to transformation from competing land uses, such as industrial agriculture and forestry, mining, logging, intensive tourism, residential development, or where there are pressures for resource use within the site itself (see Case study 4, Western Siberia). Thus, formal, legal protected area status may help custodians conserve their areas.

It is important that careful identification of sites and their indigenous local community or religious custodians be undertaken. It is also important that discussions be based on respect and trust and use appropriate negotiation techniques[3] and, where suitable, that the dialogue be effectively mediated. It is critical that all decisions be based on informed consent of appropriate and legitimate stakeholders. An increasing number of examples exist where indigenous communities are taking the lead in the development of protected areas that are rooted in their own cultural norms and recognised by governments (see Case study 15, Dhimurru IPA). A new protected area can be an appropriate tool for safeguarding sacred natural sites if it is created in a sensitive and respectful way. Guidelines by the CBD and now by IUCN/UNESCO provide a framework for best practice.

Considerable caution, however, needs to be exercised when considering the protection of sacred sites through national protected area legislation. Very few countries have developed protected area legislation that recognises sacred natural sites and their custodians in sensitive and appropriate ways. Many protected area agencies follow the protectionist model with the use of military-style enforcement methods. Unsuitable laws and inappropriate application could, as has occurred in the past, disempower minorities. Therefore, any government support and recognition to sacred natural sites should be negotiated by all parties and be made in full consultation and with the free, prior and informed consent of custodians (guidelines 1.4 and 3.1). The concept of free, prior and informed consent applies, at least in international law, primarily to indigenous communities, but in the case of negotiations with custodians of sacred natural sites it is recommended that the principles apply equally to all custodians.

Another option is to set up a multicultural process for defining the national protected area system. In this case, each cultural group has the opportunity to nominate sites for protection based on their own criteria and priorities. From experience in Canada, it has been found that while the use of scientific methods to identify areas representative of different biogeographic regions is prevalent in modern society, traditional societies tend to prefer value-based criteria that give particular weight to sacred natural sites, sites of historical importance, or sites identified with the origins of the particular cultural group.

Case study 2. Threatened first by tourism, Misali Island, Zanzibar, Tanzania is recognised as a Sacred Gift for a Living Planet

The government of Zanzibar's original plan for Misali Island, which embodies religious values to the predominantly Islamic Zanzibar community, was for a foreign-owned destination resort. To launch the tourist development plan, a concession agreement was signed. Fishermen and conservationists, on learning of the decision, banded together to lobby against the plan, and managed to persuade the government to reverse the decision. After several years of hard work, the government of Zanzibar not only recognised the fishermen as the lead managers of the Misali Island Marine Conservation Area but also offered the project to the world as a "Sacred Gift for a Living Planet".

Misali Island is a small (one-hectare) coral islet located off the west coast of Pemba Island, the northern of two main islands that make up the Zanzibar archipelago, off Tanzania. For about 26 fishing villages and the 11,500 people of Pemba, Misali, although small, is crucial to their fishing livelihoods. Misali Island's importance for fishing stems from two sources. First, the coral reefs surrounding it are rich fishing grounds, and second, it is a *dago*, that is, a temporary fishing camp. Many fishermen who use the island's waters camp out on Misali, which has no permanent habitation. When exactly, and for how long they stay, depends on which type of fishing gear they use, and on the monthly lunar/tidal cycle that is appropriate to the target species. Octopus fishermen, for example, tend to camp during neap tides when the coral reef is most accessible.

[3] E.g. principled negotiation (Fisher *et al.*, 1992).

Case study 2. Threatened first by tourism, Misali Island, Zanzibar, Tanzania is recognised as a Sacred Gift for a Living Planet (cont.)

In the early 1990s, local fishermen faced two main threats to their livelihoods. First was the leasing of Misali Island as a tourist resort, which threatened to prevent fishermen using it as a *dago*, and would have seriously disrupted their livelihoods. Second was the increasing use of destructive fishing techniques including *kigumi* (described below) and, to a lesser extent, dynamite fishing.

In response to the tourist threat, the fishermen collaborated with conservationists, who had recognised the ecological value of the island's coral reef, to challenge the government's decision to concession the island as a tourist resort to a European tour operator. After considerable lobbying, the concession decision was reversed and a proposal was then developed for a co-management area run by the fishermen, which was finally declared the Misali Island Marine Conservation Area in 1998.

The challenge remained, however, to control destructive fishing techniques, and bring real benefits to fishermen. *Kigumi* is a locally developed, but very destructive fishing method – groups of fishermen working together use sticks and poles to beat the water and coral reef to scare the fish into a large seine net. This net has a very fine inner weave – mosquito netting is often used – that removes even the smallest fry from the reef. The patch of reef left behind is physically damaged and devoid of fish.

It was during the campaign to encourage sustainable fishing that Misali Island's sacred values emerged. It is considered a holy island in local Islamic belief. According to tradition, the Prophet Hadhara once visited the island and asked the fishermen for a *msala* or prayer mat. The fishermen did not have one and so the Prophet declared he would use the island itself as his prayer mat. It was this act of praying directly on the island that is said to have given the island its name and established it as a sacred site.

Building on this local tradition and the fact that Misali's fishermen are almost exclusively Muslim, a ground-breaking programme was developed to work with the mosque Imams to bring out the strong environmental teachings of Islam in support of fisheries co-management and sustainable tourism at Misali. Named the "Misali Ethics Programme" operated by CARE Tanzania, with support from the Islamic Foundation for Ecology and Environmental Management, this may be the first time a conservation project based on Islamic ethics has been successfully implemented for marine conservation.[4]

The programme organized an initial workshop which brought together religious leaders, government officials and fishing communities, to discuss the teachings of the Qur'an and its guidance on the use of the environment. This approach was much appreciated locally[5] and it was proposed that the management of the Misali Island Marine Conservation Area should be based on the ethical principles laid down by Islam. The programme, which worked in 12 villages, held workshops for mosque leaders, developed posters and held competitions for Madrassa schools. The project has developed training materials for Imams, and supports greater understanding of religious teachings related to conservation.

Early on, both the government and the Mufti of Zanzibar supported the project and in 2000 the government offered the example of the Misali Ethics Programme as a "Sacred Gift for a Living Planet" as part of the programme of the World Wide Fund for Nature (WWF) and the Alliance of Religion and Conservation (ARC).

The programme has had positive impacts on Pemba. A baseline study at the outset of the project showed that only 34% of fishermen thought that Islam was relevant to their use of the sea and its resources. Later in the project this was found to have increased to 66%. Fishermen were practising some specific conservation measures, and lessons learnt had spread to other villages. A key output of the project was publishing and distributing a teacher's guidebook (Khalid and Thani, 2007), which has been translated into Swahili and is being distributed throughout the fishing communities of Pemba. It has a wider relevance to Islamic communities elsewhere.

Source: From Dudley *et al.*, 2005; Khalid and Thani, 2007.

[4] Khalid and Thani, 2007.
[5] Khalid and Thani, 2007.

Case study 3. Sacred land zoning in Japan

The holistic traditional land-use zoning of Japan was based on three main areas: *Okuyama,* the higher land of the sacred mountains, *Satayoma,* the lower slopes of the mountains containing sustainably managed forests, and *Hitozato*, the cleared and intensively settled farming areas. Within the latter are found numerous small sacred groves, called *Chinju-no-mori,* which contain temples and shrines and are reservoirs of local biodiversity. The Japanese people worship gods in the natural world, known as *kami*, which are believed to number some 8 million, and the *Chinju-no-mori* stand as important elements that sustain the sacred atmosphere and the concept of harmonious co-existence between nature and humanity. Recent changes in Japanese lifestyle emphasising economic success and materialism have overwhelmed and undermined this holistic view of the Japanese archipelago. In the renewed and urgent search for sustainability, it may be useful to reconsider the sacred framework established by our ancestors and symbolically represented by the *Chinju-no-mori*.

Source: Iwatsuki, 2005; 2007.

1.4 Consultation

Include the appropriate traditional cultural custodians, practitioners and leaders in all discussions and seek their consent regarding the recognition and management of sacred natural sites within or near protected areas.

During the process of recognising sacred sites that are located within legally protected areas, it is very important to consult with the appropriate traditional custodians. This is an essential first step in engaging with custodians over management of protected areas, and may lead to higher-level participation and co-management. Consultation must start early, be clearly defined, and continue for the long term. A liaison person on the protected area staff should be in charge of communication, and all communication should be handled carefully. Funding should ideally be made available to pay for travel to meetings, and to cover the expenses of custodians involved in recognising and managing sacred natural sites. Determining the appropriate custodians, practitioners and leaders for consultation is a complex and delicate task, requiring care and time (see Box 7 on stakeholder analysis, and guidelines 3.3 and 3.4). "Custodian" does not necessarily imply a current geographic connection with the site and this needs to be taken into consideration.

1.5 Holistic models

Recognise that sacred natural sites integrate social, cultural, environmental and economic values into holistic management models that are part of the tangible and intangible heritage of humankind.

Many sacred natural sites have been managed under traditional systems for centuries, and in some cases for millennia. These represent some of the most successful management systems in existence and are original models of a sustainable relationship between humans and nature (e.g. Case studies 3, Japan; 15, Sri Pada-Adams Peak; 10, Ka'sankwa; 14, Rila). As such, their conservation as complete nature-human systems represents an important achievement for humanity.

These systems are in contrast to much of the "development" that has dominated the post-World War II period and which continues to be widely perceived as synonymous with Western-style modernization.[6] Local and traditional cultures have largely been seen to be a brake on development. Yet the current, profound ecological crisis has renewed interest in concepts both of sustainability and resilience. "Ours is a period when the human community is in search of new and sustaining relationships to earth amidst an environmental crisis that threatens the very existence of all life forms on the planet".[7]

It would be ideal to consider the development of "multicultural protected area systems" that fully recognise areas of importance as identified by the criteria used by each local culture. With the exception of Canada, there are currently few examples of this, but the Rila Monastery Natural Park in Bulgaria fully recognises the spiritual and natural values of the area, and maintains the connections between these values as a management objective (Case study 14).

2 Integrate sacred natural sites located in protected areas into planning processes and management programmes.

2.1 Park planning

Initiate planning processes to revise management plans to include the management of sacred natural sites located inside protected area boundaries.

Protected area management planning is one of the most widely accepted tools of site-based conservation. Since the early 1990s

[6] Eade, 2002.
[7] Tucker and Grim, 2001.

Box 5. Cultural planning in the US Park Service

US National Park Service Policy on Cultural Resources

Planning

Effective park stewardship requires informed decision making about a park's cultural resources. This is best accomplished through a comprehensive planning process. Effective planning is based on an understanding of what a park's cultural resources are, and why those resources are significant. To gain this understanding, the Service must obtain baseline data on the nature and types of cultural resources, and their (1) distribution; (2) condition; (3) significance; and (4) local, regional and national contexts. Cultural resource planning, and the resource evaluation process that is part of it, will include consultation with cultural resource specialists and scholars having relevant expertise; traditionally associated peoples; and other stakeholders. Current scholarship and needs for research are considered in this process, along with the park's legislative history and other relevant information.

Planning decisions will follow analysis of how proposals might affect the values that make resources significant, and the consideration of alternatives that might avoid or mitigate potential adverse effects. Planning will always seek to avoid harm to cultural resources, and consider the values of traditionally associated groups. To ensure that approaches and alternatives for resource preservation have been identified and considered, planning processes that could affect cultural resources must include cultural resource specialists, traditionally associated peoples, and other stakeholders, and provide them with appropriate notification about opportunities to become involved ...

Many cultural landscapes are significant because of their historic land use and practices. When land use is a primary reason for the significance of a landscape, the objective of treatment will be to balance the perpetuation of use with the retention of the tangible evidence that represents its history. The variety and arrangement of cultural and natural features in a landscape often have sacred or other continuing importance in the ethnic histories and cultural vigor of associated peoples. These features and their past and present-day uses will be identified, and the beliefs, attitudes, practices, traditions, and values of traditionally associated peoples will be considered in any planning decisions.

Source: U.S. National Park Service, 2001: http://www.nps.gov/refdesk/mp/chapter5.htm

there has been increasing emphasis on the process of management planning as an important way to ensure that the final plan has wide agreement and support. Participation of key stakeholders is a critical element and has become standard best practice for park planning (see also guidelines 3.1, 3.2 and 3.3).[8] The ecosystem approach (guideline 2.7) adopted by the Convention on Biological Diversity further establishes that planning should be carried out at the level within the institutional hierarchy that is closest to the resources in question. Spiritual values and sacred natural sites, however, have generally not been included in protected area planning. In all future management plans, cultural and spiritual values and the protection and management of sacred natural sites should be considered. It would be ideal to incorporate or, where possible, base planning and management on the traditional management frameworks of custodians. The United States National Park Service, representing one approach, has developed a planning policy for cultural resources, an extract of which can be found in Box 5. For a comprehensive example of taking cultural elements into account in Guatemala, see Annex 1.

2.2 Identify sacred natural sites

Where secrecy is not an issue and in close collaboration and respecting the rights of traditional custodians, identify the location, nature, use and governance arrangements of sacred sites within and around protected areas as part of a participatory management planning process.

With full respect to prior consent (guideline 3.1), confidentiality (guideline 2.3) and with the local custodians' support and participation, a process can be undertaken to identify sacred sites within the protected area and establish the nature of their institutional management, governance structures and traditional regulations for management.

A useful approach is the joint assessment of the customary use of resources through participatory mapping and documentation, as well as the use of participatory geographic information systems (PGIS).[9]

It will be important to gain an understanding of the traditional decision-making and governance arrangements, which can show considerable variation from place to place. Custodian institutions have typically developed detailed rules regarding all aspects of the management of sacred natural sites and a common

[8] Thomas and Middleton, 2003.

[9] Rambaldi *et al.*, 2006; see Annex 8.

Case study 4. Specially protected areas and species of Western Siberia, Russian Federation

The longstanding relationship between the Khanty and Mansy indigenous communities and their land has not only resulted in the overall protection of the landscape that contains a large number of areas considered sacred, but also in the survival of endangered wildlife, notably the Eurasian Beaver (*Castor fiber*). The survival of the beaver can, in fact, be almost entirely attributed to the sacred status it is accorded by these communities. In the rest of the region the beaver was hunted to extinction by the 17th century. One of the sacred natural sites where beavers were protected by local communities was later recognised as the Malaya Sosva Natural Reserve.

Located in the Khanty-Mansiysk Autonomous Okrug – Ugra of Western Siberia, these sacred sites are now threatened by oil exploitation. The difficulty of sustainably protecting these sites is compounded by the fact that for the traditional custodians the undisturbed nature of the sites is of paramount importance. The mere act of an unauthorized visit can strip them of their sanctity and lead to abandonment.

Working with the indigenous communities, the conservation authorities of the Autonomous Okrug have established 26 Specially Protected Natural Areas with a total area of 3.7 million hectares (accounting for 6.3 % of the Okrug's area). This area contains 400 sacred natural sites, the exact locations of which remain undisclosed. In this way, it is hoped that the impacts of oil extraction will be mitigated, and that biodiversity and local cultural and spiritual values, as well as their relevance to the indigenous communities, will be maintained.

Source: Merkushina, 2007.

feature of many sacred sites is their restricted access and use. Wherever possible, respect and support custodians' own regulations regarding the management of sacred natural sites, reflecting these in protected area rules and policies as appropriate.

2.3 Respect confidentiality

Ensure that pressure is not exerted on custodians to reveal the location or other information about sacred natural sites and, whenever requested, establish mechanisms to safeguard confidential information shared with protected area agencies.

While the identification of sacred natural sites within protected areas is useful for protected area management, no pressure should be exerted on local communities to reveal the location of their sacred natural sites, nor the details of their cultural values, practices, history or use. In some instances, sacred natural sites are concealed or access is restricted to a specific gender or age group from within the custodian community and their existence cannot be revealed to the uninitiated. Likewise, the mere presence of outsiders at a sacred site can reduce or even terminate its sacred value and cause it to be abandoned.

One approach that managers have taken when custodians do not wish to divulge the location of sacred natural sites within a protected area is to set up a mechanism to inform the custodians of the intentions for management. Feedback is then requested if these plans conflict with the sacred sites. Based on the feedback, management programmes can be adjusted. In this way, protected area managers can be responsive to the needs of custodians, without having to know the *actual location of the site or details regarding its use.*

When confidential cultural information is shared, all necessary means should be taken to ensure that confidentiality is maintained and to prevent disclosure to unauthorized people and the public. This can be achieved in a number of ways. A system could be agreed, for example, with custodians, in which an accepted and trusted member of the protected area staff works with the community and becomes a holder of necessary information. If limited documentation is allowed, this should be kept in a secure way.

2.4 Demarcate or conceal

Where appropriate and to enhance protection, either clearly demarcate specific sacred natural sites, or alternatively, to respect the need for secrecy, locate sacred natural sites within larger strictly protected zones so exact locations remain confidential.

To enhance the protection of sacred natural sites a variety of strategies may be required. The decision as to the most appropriate approach will be specific not only to different locations, but also over time. In some cases, the best approach will be to increase the knowledge, awareness and even publicity about sites, a component of which may be the clear demarcation of the boundaries. In other cases, where risks of damage and even desecration may arise out of greater public knowledge, sites will be best protected if they are little known and even concealed. Designating larger areas as strictly protected zones is one approach to achieve this that can be used in low population density areas.

Other methods to protect sacred natural sites include appropriate use of zoning, the layout of visitor trails and roads to avoid sensitive areas, the use and careful management of permits for wilderness areas, and the compulsory use of authorized guides. In situations of high visitor pressure, clear signage and, in extreme cases, the fencing off of particularly sensitive areas may be required (see photo plates).

2.5 Zoning

Establish support, buffer and transition zones around and near sacred sites, especially those that are vulnerable to adverse external impacts.

Zoning is a standard tool of land-use planning and management, both within protected area management plans, and also in the land-use master plans of development planning authorities. Zoning can be used in a multitude of ways to support the conservation of sacred natural sites, by applying specific management objectives to particular geographical areas, and by controlling use and access.

Many sacred natural sites have traditional zoning arrangements that are analogous to protected area zoning. For example, they often contain a core area or "inner sanctum" where only spiritual leaders are allowed, as well as other areas for ceremonial use or access. Thus, new zoning of sacred natural sites that occur within protected areas will ideally build upon the traditional zoning established by custodians, often over centuries of practice. Efforts should be undertaken by protected area managers to properly understand any traditional zoning that may exist.

Zoning can also be used to achieve the objectives of controlled access or concealment discussed in guideline 2.4. Sacred natural sites that require minimal disturbance, for example, can have larger zones surrounding them into which access can be limited.

Areas within a sacred natural site can also be zoned to enhance management. At Kaya Kinondo, one of the sacred forests that make up the Kenya Kaya World Heritage Site, the Kaya elders have built upon the traditional internal zoning structure to allow for ecotourism. They have developed specific visitor trails and a visitor zone, dress codes for access, as well as areas in which photography is, or is not, allowed (see Case study 5, and photo plates).

The UNESCO Biosphere Reserve concept can also be effectively applied to assist with the conservation of sacred natural sites. In this model, the "core" area could be the sacred natural site itself, and buffer and transition zones could be created around the core area.[10] In certain situations, especially where land surrounding the sacred natural site has become modified, the buffer zone could support other livelihood or conservation objectives. This can support the incomes of community members. An example of this is the use of economically important trees (e.g. fruit or fuel wood) at the periphery of a sacred grove. This may help with the grove's protection by clearly marking its location while also supporting local livelihoods. In other cases, a portion of the buffer zone could consist of the native species found at the sacred natural site. This will support the biodiversity values of the area and habitat restoration as well (guideline 2.6).

Case study 5. Kaya sacred forests, Kenya

The discovery by plant ecologists in the late 1980s that the small, but biologically outstanding patches of Kenyan coastal forests were, in fact, community-managed sacred groves, sparked a productive collaboration with the traditional custodians, the Mijikenda. In 2007, this collaboration resulted in the Kaya forests being nominated for inscription on the World Heritage List.

Kaya means "homestead" and the Kayas represent the original village sites of the nine Mijikenda tribes. Threatened by other groups, the Mijikenda sought refuge in the dense coastal forests but later moved out of the Kayas and cultivated and cleared land outside. Forty-nine Kaya forests remain as small but biodiverse remnants of the once extensive forests. Research has shown that while the Kayas now represent only 10% of the remaining coastal forests in Kenya, they contain a disproportionately large number of rare plants. Seven out of 27 forest sites in the coastal area with highest relative conservation values were Kaya forests. More than half of Kenya's rare plants are found in the coast region and many of these occur in the Kaya forests.

Threats: Since the 1970s, development pressures have meant that portions of some Kaya forests have been lost directly to tourist development, a pressure that continues to grow. Other threats include unauthorized harvesting of trees for tourist wood carvings, as well as pressure to meet the livelihood needs of local people. Although the Mijikenda themselves respect traditional regulations, the strength of traditional institutions has been eroded, particularly through movement into the area of people less willing to submit to Mijikenda laws. Population growth, livelihood expectations and poverty have added to the pressures. The

[10] UNESCO, 2003b.

development of high-end tourism on Kenya's coast has led to a classic divide between the developed hotel strip and the underdeveloped rural hinterland where most of the Kayas are located.[11]

Recognising the threats to their sacred forests, the Kaya elders lobbied for government support. Joint conservation work in the 1980s resulted in the establishment of the Coastal Forest Conservation Unit in 1992. This was a project of the National Museums of Kenya with support from WWF in partnership with the Kaya elders.

Legal protection: In the process of gaining legal protection, the best mechanism was a subject of debate. Specifically, the issues under consideration were: the strong spiritual, cultural and historic values of the Kayas; a relatively high but controlled level of community access; and the need for strict protection. How to protect but allow for traditional use? Options considered were protection as Forest Reserves, National Monuments or even as National Parks. Each option offered a different lead agency, each with its own set of legislation, and use and management structure, and each with positive and negative implications.

National Monument status, managed by the National Museums of Kenya, was selected as the most suitable option. This gave the best balance of strict protection, local access and use, and active protection of cultural and biological values. Even then, the relevant heritage and museums legislation needed to be revised to accommodate the Kayas. Subsequently, most of the Kayas were gazetted as National Monuments.

Despite these modifications, it has been difficult to formally accommodate the traditional laws of the Mijikenda elders, as these primarily apply only to the Mijikenda people and belong to oral tradition not recognised by national law. Despite this, there is an operational understanding of the importance and role of traditional management. The elders carry out daily site management and enforce the traditional regulations and, in some cases, employ Kaya guards. The elders call in staff of the National Museums of Kenya, who maintain a regional presence, only for those cases they cannot deal with themselves.

Cultural conservation: Conservation programmes have led to some revival of Kaya management and traditional ceremonies. There has been a strengthening of the Council of Elders at each Kaya, and, for example, in Kwale District, a district-wide meeting of elders is held regularly, and the local Chairman is an ex-officio member of the District Council. In this way, the traditional leadership is able to represent the interests of the Kaya elders at a political level.

Ecotourism: Efforts are now underway to increase revenue for management. At one Kaya, Kaya Kinondo in Kwale District, WWF has supported an ecotourism project. The elders chair the management committee and have identified the areas of the Kaya that can receive visitors and where photos can be taken, as well as areas that are off limits and where no photography is allowed. They also set out conditions by which, for example, trails could be constructed. When clearing one area of trail the elders felt that too much vegetation had been cleared and fined the project as a result. A modest visitor centre has been established on the edge of the Kaya and volunteer guides drawn from the local community have received training. The guides run tour desks in nearby hotels promoting half-day excursions, which include a village visit and meal.

Despite the elders zoning a small portion of the Kaya for visitor access, they still felt that nature periodically needed a complete rest from people. Therefore, they decided that every fourth day, the rest day in the Mijikenda traditional four day week, the forest should be closed to tourism. A translation of the Mijikenda calendar to the Western calendar identifies for tourists which days the forest is closed. Exceptions to this rule can be requested from the elders on a case-by-case basis.

Long-term survival of the Kayas: In the last 20 years, great strides have been taken to secure the long-term future of the Kayas. Their high biodiversity values have been proven, and traditional management is being acknowledged and revitalized. They have legal recognition, and direct threats from land-use change and overuse have been contained if not halted. Despite this, the government is providing little direct financing, the National Museums of Kenya remain dependent on outside support, and Kaya elders are left largely to their own resources for day-to-day management. This presents significant challenges, scattered as the forests and elders are over a large area.

In the opinion of people working in the area, the survival of the Kaya forests over the long term depends on improved livelihoods and income generation for the communities around the Kayas. If livelihoods improve and alternative income generation can be provided, the existing forests will survive. A microfinance scheme has been initiated to support economic development at three Kayas in Kwale District. However, this is in the early stages of development and is not yet having a major impact on livelihoods (Wild *et al.*, 2007).

Author: Written by R. Wild, based on interviews with Elder Mr Abdalla Ali Mnyensi, Chairman of Kaya Kinondo Council of Elders, Chairman of Kwale District Kaya Committee, and nominated Kaya District Councillor. Elders Omar Rashid Kituzo, Kaya Kinondo Elder, and Deputy Chairman of Ecotourism Management Committee, Ali Mwalimu Mwarandani Kaya Forest Guard, Hemed Mwafundjo Coordinator, Kaya Kinondo Ecotourism Project. Also referenced Githitho, 2003 and 2006.

[11] Sindiga, 1996.

2.6 Linkages and restoration

Create ecological corridors between sacred natural sites and other suitable areas of similar ecology for connectivity, and in degraded landscapes consider restoring sacred natural sites as an important initial step to reviving a wider area.

Sacred natural sites are often the last refuge of rare, endemic or threatened species in heavily altered and degraded landscapes. Frequently this is due to the protection afforded to them by traditional custodians. Due to their biodiversity values they can form the core areas from which efforts to restore larger areas of natural vegetation can be initiated. For example, the great plant diversity contained within sacred groves could be used as a resource for ecosystem rehabilitation based on successional concepts.[12] Ideally, the recognition or establishment of connecting corridors of similar or related vegetation between sacred natural sites will strengthen their continuity and conservation value. Sacred natural sites may also play a role as ecological "stepping stones" or areas of habitat that allow species to migrate within a landscape. They may prove to be important for species survival during climate change disruptions. Further research is required, however, to understand the role of sacred natural sites in ecological networks across a landscape in relation to the effects of climate change (guideline 2.8).

2.7 Ecosystem approach

Adopt the ecosystem approach as the key strategy for the integrated management of land, water and living resources that promotes conservation and sustainable use in an equitable way and also includes cultural and spiritual values.

The ecosystem approach has in recent years become a key part of the Convention on Biological Diversity.[13] It calls upon signatory nations to develop practical expressions for national policies and legislation, as well as appropriate implementation and efforts to promote the sharing of experience. The IUCN Commission on Ecosystem Management (CEM) has promoted the ecosystem approach and has developed a procedure to put its principles into practice.[14] Both the approach and this procedure are briefly described below.

The ecosystem approach is a strategy for the integrated management of land, water and living resources that promotes conservation and sustainable use in an equitable way and helps reach a balance of the three objectives of the Convention: biodiversity conservation, its sustainable use, and the fair and equitable sharing of its costs and benefits.

The procedure to put the principles of the ecosystem approach into practice organizes 12 principles into five sequenced steps:[15]

- The first and biggest issues: area and key stakeholders.
- The next set of issues: ecosystem structure, function, health and management.
- Economic issues.
- Adaptive management over space: impact on and from adjacent ecosystems.
- Adaptive management over time: long-term goals, flexible ways of reaching them.

The ecosystem approach calls for the identification of stakeholders at an early stage and sets out four main steps:

Step 1. Identify the key stakeholders with interests in the proposed ecosystem.

Step 2. Weight them as primary, secondary or tertiary stakeholders, and assess their views in that light.

Those who are most dependent upon the resources usually must be strongly, positively weighted as primary.

Over-powerful voices which may need to be weighted as secondary or tertiary may include those who live near the resource but do not greatly depend on it, e.g. government officials, and international conservation organizations.

Step 3. Assess the stakeholders' relative management capacity and commitment in regard to the ecosystem.

Step 4. Set up a stakeholder forum which will meet regularly.

The ecosystem approach provides an effective and widely accepted framework for land management, and one in which sacred natural sites can be addressed. Many resources are now available for the implementation of the ecosystem approach.[16] Particularly important for sacred natural sites is the understanding of traditional ecological knowledge and indigenous governance systems (guideline 4.3).

---

12 Ramakrishnan *et al.*,1994, 1996, in Ramakrishnan, 2003.
13 It was adopted by the Convention at its 5th Conference of Parties, in the year 2000 (Decision V/6).
14 Shepherd, 2003; 2004.
15 Shepherd, 2003; 2004.
16 http://www.iucn.org/themes/cem/ourwork/ecapproach/index.html

2.8 Landscape approach

Take a landscape approach to sacred natural sites, recognising their role in wider cultural landscapes, protected area systems, ecological corridors and other land uses.

Sacred natural sites, in addition to their own values, form part of wider landscapes. In some communities, the whole landscape is sacred, with some areas having specific functions and even higher spiritual values. In some sacred landscapes, legal protected areas may form just one small part of the total area. For other communities, sacred natural sites are located in more "ordinary" landscapes but may be linked through a range of ways: mythically, historically, or physically by pilgrim routes or common management. Biologically, many such landscapes are modified by human activity over time. It is increasingly recognised that habitat modification by people, at least at some level, appears to be the rule rather than the exception, with even the most remote areas showing modifications by indigenous and local people. Thus, over generations many landscapes have evolved in which humanity has had significant influence and has modified vegetation types. Within such landscapes are often located special sacred sites largely of natural origin.

A mechanism to conserve these landscapes has become known as the "protected landscape approach". The protected landscape approach links conservation of nature and culture and sustains people's relationship to the land while fostering their stewardship of it.[17]

As described by Adrian Phillips:

"Landscape can be seen as a meeting ground between:

- Nature and people – and how these have interacted to create a distinct ***place***;
- Past and present – and how therefore landscape provides a ***record*** of our natural and cultural history;
- Tangible and intangible values – and how these come together in the landscape to give us a sense of ***identity***."[18]

Sacred natural sites of all kinds have played a key role in the creation of distinct places that record natural and cultural histories and provide a sense of identity. The protected landscape approach is therefore important to the conservation of sacred natural sites.

In terms of conservation history, the protection of such landscapes has received less emphasis than the so-called "strictly protected areas" usually considered as IUCN Categories I–IV. Approaches to conservation that recognise the value of landscapes which contain both natural and human elements have, however, been gaining ground in recent years.[19] The two main formal mechanisms for recognition are the IUCN Category V – Protected Landscapes/Seascapes and the World Heritage Convention's Cultural Landscapes.[20] These two complementary approaches have different emphases. In Category V – Protected Landscapes/Seascapes, the primary emphasis is on the natural environment, biodiversity and ecosystem integrity, mostly at the national or sub-national level. In the Cultural Landscapes of the World Heritage Convention, designation has been based on human history, the continuity of cultural traditions, social values and aspirations of outstanding universal value.[21]

2.9 Support development planning recognition

Development planning authorities are the main planners of land use in areas outside many protected area systems. Seek their and other stakeholders' support for the recognition of sacred natural sites in the wider countryside.

While protected area managers are the main focus of these guidelines, development planning authorities play a crucial role in decisions over land use. Development planners are responsible for drawing up regional plans as well as more local-level strategic development or land-use plans. Protected area agencies are often consulted by development planners and can make comments on the development planning process. They regularly advise planning authorities as to the biological and cultural values of areas of land, and can advocate for the recognition and protection of sacred sites in development planning.

Planning authorities are also likely to be involved in the planning of new protected areas. They often have legal requirements to consult with local interests during the development of new plans and for new developments. In many countries, development planning procedures are not, however, well developed and planners often lack detailed ecological and cultural knowledge upon which to base planning decisions. Protected area agencies may be in a position to support regional planners with information related to the cultural and biological values of particular areas, and provide support for working with sacred site custodians.

17 Brown *et al.*, 2005.
18 Phillips, 2005, original emphasis.
19 Brown *et al.*, 2005.
20 Phillips, 2005.
21 Brown *et al.*, 2005.

2.10 Protected area categories and governance

Recognise that sacred natural sites exist in all of the IUCN protected area categories and governance types, and that those that fall outside formal protected area systems can be recognised and supported through different legal and traditional mechanisms according to the desires of their custodians, including as community conserved areas when appropriate.

Protected area categories. Sacred natural sites occur within all IUCN categories of protected area.[22] Some sacred natural sites that have been designated specifically within protected area systems have been recognised as Category III – Natural Monuments. However, analysis shows that any of the six IUCN categories may, in fact, be useful for specific cases. Borrini-Feyerabend *et al.* (2004) note that "an analysis of specific cases will help to determine if strict protection objectives can be compatible with community involvement in conservation, and to evaluate the conservation effectiveness of traditional practices, including area protection and resource-use restriction imposed by communities themselves". In fact, it is likely that many sacred natural sites, in terms of access and use, are analogous to Category Ia – Strict Nature Reserves, the most highly protected IUCN category. For example, many of the sacred groves in Africa and Asia have strict controls on all but the most limited extractive use, with strong prohibitions on harvesting, hunting and access. It is for this reason that these groves retain very important biodiversity values.

Another example of restricted use is the petition to the Chinese government to ban climbing of Mt Kawagebo, based on the sacred values of local people. Mt Kawagebo, in Yunnan Province, the very sacred Mount Kailas in the Tibet Autonomous Region, Machhapuchhare in Nepal, and Gangkhar Puensum in Bhutan are four of the few remaining Himalayan peaks "unconquered" by mountain climbers (see Box 6).

Protected area governance. Recently, increased attention has been paid to the plurality of governance types of protected areas, particularly regarding community-level governance. At the World Parks Congress in 2003, the classification of "Community Conserved Areas" (CCA) for sacred natural areas and other sites that have been protected over long periods by local communities and indigenous people was introduced. This was further articulated in WCPA Best Practice Protected Area Guidelines Series No. 11, on indigenous and local communities and protected areas.[23] Many of the recommendations and policy options identified in that volume have great relevance to sacred natural sites and support the guidelines set forth in this volume (see also Table 1 and Annex 8).

2.11 International dimension

Recognise that some sacred natural sites, and the cultures that hold them sacred, cross international boundaries and that some may be within or may surround existing or potential transboundary parks for peace.

Protected areas have played a role in improved cooperation across national boundaries through transboundary protected areas. The concept of cooperation between parks across national boundaries started as early as 1932 in Canada and the United States. Other transboundary protected areas have been under development in recent years.[24] A specific category of transboundary protected area is the "park for peace" which has the specific objective of the

Box 6. Sacred mountains off limits to mountaineers

Machhapuchhare in Nepal is considered sacred by the Gurung people, and has never been climbed to its summit. The first and only known attempted ascent of the mountain was in 1957.[25] The expedition climbed to within 50m of the summit via the North Ridge, but did not complete the ascent; they had promised local elders not to set foot on the actual summit out of respect, and had to leave offerings at a shrine in the valley below to pacify the mountain's goddess. Since then, Machhapuchhare has been officially declared sacred, and it is out of bounds to climbers.

Gangkhar Puensum is the highest mountain in Bhutan and the highest unclimbed mountain in the world, with an elevation of 7,570m and a prominence of over 2,990m. After Bhutan was opened for mountaineering in 1983, four expeditions made failed summit attempts on Gangkhar Puensum in 1985 and 1986. However, in 1999, a team successfully climbed a subsidiary peak of the mountain, gaining access from Tibet. Since 1994, climbing of mountains higher than 6,000m has been prohibited in Bhutan out of respect for local spiritual beliefs, and since 2003, mountaineering has been forbidden completely.

Mount Kailas is the most sacred mountain in the world for more than a billion people – followers of Hinduism, Buddhism, Jainism and the indigenous Tibetan tradition of Bon. According to Buddhist tradition, Tibet's most famous yogi, Milarepa, flew magically to the summit in the 12th century and there have been no recorded attempts to climb Mt Kailas since.

Source: http://en.wikipedia.org/wiki

[22] Verschuuren *et al.*, 2007.
[23] Borrini-Feyerabend *et al.*, 2004.
[24] Sandwith *et al.*, 2001.
[25] Reynolds, 2003.

promotion of peace and cooperation, as well as the protection and maintenance of natural and cultural values.

A number of transboundary protected areas have cultural and spiritual values and some are sacred natural sites. The Kgalagadi Transfrontier Park between Botswana and South Africa, for example, has long had deep sacred values for the Khoi and San peoples. Currently, transfrontier cooperation is also ongoing between the Russian Federation and Mongolia, who are carrying out joint research over the water catchment of Lake Baikal, sacred to the Buryat people.[26]

"Our sacred Lake Baikal (between Mongolia and Russia) – the oldest and deepest lake in the world – is home to more than 3,000 species, almost half of which are endemic to the area. Buryat clans connect their origins directly with Baikal and trace their lineage to natural forefathers – the animals and trees of the lake. This high degree of diversity of life forms in sacred sites could be explained by the fact that sacred sites are places of origin of certain families or clans and have been protected because many of the forms are believed to be our ancestors".[27]

Furthering understanding of the cultural and spiritual values of transboundary protected areas is valuable. At the same time, sacred natural sites can be places where other important values, such as peace building, can receive greater attention.

3 Promote stakeholder consent, participation, inclusion and collaboration.

3.1 Prior consent

Ascertain the free, prior and informed consent of appropriate custodians before including sacred natural sites within new formal protected areas and protected area systems and when developing management policies affecting sacred places.

Free, prior and informed consent has become a widely accepted principle.

"The right of peoples to give or withhold their free, prior and informed consent (known as FPIC) to actions that affect their lands, territories and natural resources has become widely recognised and come to be seen as especially important for indigenous peoples in their dealings with non-state actors seeking to control, or to gain access to, their lands and resources whether for development or conservation".[28]

Free, prior and informed consent is established in international law and accepted by a wide group of international institutions, including the CBD and IUCN.[29] The principle is less well reflected in national laws and significant gaps occur in its implementation.[30] As elaborated by the International Fund for Agricultural Development,[31] "Free, prior and informed consent recognizes indigenous peoples' inherent and prior rights to their lands and resources and respects their legitimate authority to require that third parties enter into an equal and respectful relationship with them, based on the principle of informed consent".[32] The underlying principles of free, prior and informed consent can be summarised as follows: (i) information about and consultation on any proposed initiative and its likely impacts; (ii) meaningful participation of indigenous peoples; and (iii) representative institutions.

Indigenous peoples' right to free, prior and informed consent is also embraced in the UN Declaration on the Rights of Indigenous Peoples. Article 30 of the Declaration provides that "Indigenous peoples have the right to determine and develop priorities and strategies for the development or use of their lands, territories and other resources, including the right to require that states obtain their free and informed consent prior to the approval of any project affecting their lands, territories and other resources, particularly in connection with the development, utilization or exploitation of mineral, water or other resources".[33]

Respect for the right of indigenous people to FPIC has come to be appreciated as a crucial tool in the achievement of social and environmental sustainability. Respect for this right by conservation agencies should bring to an end the unfortunate conflicts that have developed between indigenous people and those seeking to establish protected areas in their territories.[34]

Despite the establishment of this essential right, experience indicates that there remains a considerable gap between principle and practice. To effectively use FPIC, other factors need to be brought to play to support meaningful consent, including community capacity building, good legal advice and enhancing

26 Drobyshev Yu *et al.*, 2007.
27 Khamaganova, 2007.
28 Colchester and Ferrari, 2007.
29 E.g. at the Vth World Parks Congress (Durban Accord Annex 6).
30 Colchester and Ferrari, 2007.
31 IFAD, 2005.
32 Commission on Human Rights, Sub-Commission on the Promotion and Protection of Human Rights, Working Group on Indigenous Populations, Twenty-second session, 19–23 July 2004, p.5.
33 FPP, 2004, Annex 3.
34 Colchester and Ferrari, 2007.

representativeness of indigenous institutions.[35] This often requires providing support to communities to adequately prepare for meaningful negotiations.[36]

3.2 Voluntary participation

Ensure that state or other stakeholder involvement in the management of sacred natural sites is with the consent and voluntary participation of appropriate custodians.

The voluntary participation of local people in conserving sacred natural sites is a fundamental principle. It is important that discussions be held between appropriate custodians and protected area agencies regarding programmes to support the conservation of sacred natural sites located within protected areas. The relationship between government agencies and custodians historically may not be good and trust may be lacking. Programmes for support of sacred natural sites should only be initiated with the consent of the custodians, who may fear damage or desecration to the site by outside involvement. If such fears or concerns exist, they should be fully respected and no pressure should be exerted.

Figure 4. Local adaptations of the Millennium Ecosystem Assessment conceptual framework

The Millennium Ecosystem Assessment (MA) framework was applied in a wide range of assessments at multiple scales. Particularly for the more local assessments, the framework needed to be adapted to better reflect the needs and concerns of local communities. In the case of an assessment conducted by and for indigenous communities in the Vilcanota region of Peru (see Case study 16), the framework had to be recreated from a base with the Quechua understanding of ecological and social relationships. Within the Quechua vision of the cosmos, concepts such as reciprocity (Ayni), the inseparability of space and time, and the cyclical nature of all processes (Pachakuti) are important components of the Inca definition of ecosystems. Love (Munay) and working (Llankay) bring humans to a higher state of knowledge (Yachay) about their surroundings and are therefore key concepts linking Quechua communities to the natural world. Ayllu represents the governing institutions that regulate interactions between all living beings.

The resulting framework has similarities with the MA Conceptual Framework, but the divergent features are considered to be important to the Quechua people conducting the assessment. The Vilcanota conceptual framework also includes multiple scales (Kaypacha, Hananpacha, Ukupacha); however, these represent both spatial scales and the cyclical relationship between the past, present, and future. Inherent in this concept of space and time is the adaptive capacity of the Quechua people, who welcome change and have become resilient to it through an adaptive learning process. (It is recognised that current rates of change may prove challenging to the adaptive capacities of the communities). The cross shape of the Vilcanota framework diagram represents the "Chakana," the most recognised and sacred shape to Quechua people, and orders the world through deliberative and collective decision making that emphasises reciprocity (Ayni). Pachamama is similar to a combination of the "ecosystem goods and services" and "human well-being" components of the MA framework. Pachakuti is similar to the MA "drivers" (both direct and indirect). Ayllu (and Munay, Yachay and Llankay) may be seen as responses and are more organically integrated into the cyclic process of change and adaptation.

In the Vilcanota assessment, the Quechua communities directed their work process to assess the conditions and trends of certain aspects of the Pachamama (focusing on water, soil and agrobiodiversity), how these goods and services are changing, the reasons behind the changes, the effects on the other elements of the Pachamama, how the communities have adapted and are adapting to the changes, and the state of resilience of the Quechua principles and institutions for dealing with these changes in the future. Developing the local conceptual framework from a base of local concepts and principles, as opposed to simply translating the MA framework into local terms, has allowed local communities to take ownership of their assessment process and given them the power both to assess the local environment and human populations using their own knowledge and principles of well-being and to seek responses to problems within their own cultural and spiritual institutions.

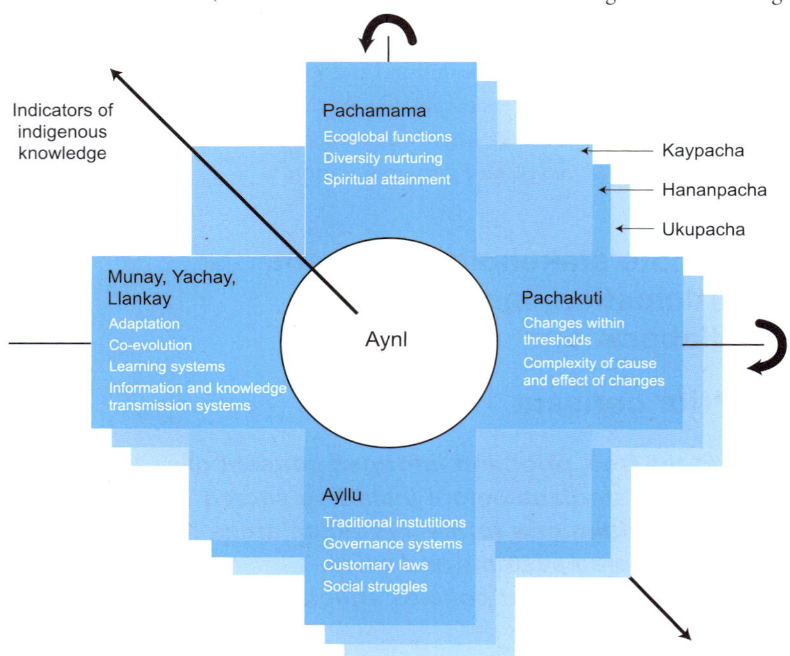

Source: Millennium Ecosystem Assessment: Vilcanota Sub-global Assessment

[35] IFAD, 2005.
[36] Colchester and Ferrari, 2007.

3.3 Inclusion

Make all efforts to ensure the full inclusion of all relevant custodians and key stakeholders, including marginalized parties, in decision making about sacred natural sites, and carefully define the processes for such decision making, including those related to higher level and national level policies.

The inclusion of a wide range of appropriate stakeholders in decision making is essential. At a site sacred to indigenous or local communities this range of stakeholders might include community heads, clan elders, chiefs, spiritual leaders, traditional owners and practitioners, knowledge holders, as well as the community at large. At the sacred site of a mainstream faith it may include various levels of the monastic leadership or religious hierarchy. A stakeholder analysis may be useful in understanding different actors that are involved in a site (Box 7).

"Participation, however, requires longer implementation periods, the flexible design of interventions, close monitoring and evaluation, and an effective system for communication amongst stakeholders".[37]

Ideally, marginalized and/or minority members of the community or communities will be included. This can present a challenge, particularly when dominant structures prevent participation based on gender or ethnicity. At the same time, there is the need to be respectful of local cultural norms, consistent with basic human rights.

The participation of a wide range of stakeholders at the community level is now accepted standard practice, although it is in need of ongoing reinforcement and sometime re-learning. Community participation in the development of national-level policies, however, is less common. The result is that policies are often developed remotely, with little knowledge of the actual situation on the ground and such policies are often inappropriate. In developing national policies for sacred natural sites it is important that community-level input is included and participation is carefully planned. It is useful to distinguish between individuals who formally "represent" communities from those who are "representative of" communities. Ideally, individuals who formally represent key stakeholders will be involved, and are mandated to speak **for** their communities. It is not always possible to achieve this, and a second option is to seek to involve individuals who are representative of key stakeholders, and can speak **about** their communities. While the former is much stronger the latter can help to better inform the policy process.

Box 7. Stakeholder identification and analysis

Stakeholder analysis has become a well accepted tool of participatory development processes and is used to understand patterns of interaction, improve interventions, establish which actors should be party to decision making, improve policy formulation and understand conflict.

Stakeholders are defined as natural resource users and managers, while stakeholder analysis refers to tools for the identification and description of stakeholders based on their attributes, interrelationships and interests, and related to a given resource.[38]

The following is a flexible set of steps for conducting stakeholder analysis:[39]

- Identify the main purpose of the analysis;
- Develop an understanding of the system and decision makers in the system;
- Identify principal stakeholders;
- Investigate stakeholder interests, characteristics and circumstances;
- Identify patterns and contexts of interaction between stakeholders; and
- Define options for management.

37 IFAD, 2005.
38 Ramírez, 1999.
39 Grimble *et al.* 1995.

Box 7. Stakeholder identification and analysis (cont.)

It is important to reflect upon the following questions:

- Who decides on the purpose of the analysis?
- Who counts most?

For the convener, these questions relate to the power, legitimacy or resources to convene others, the power to choose the criteria for inclusion or exclusion of other stakeholders, and the authority to define the reason or theme around which stakeholder analysis takes place. For other stakeholders it has to do with "being noticed" or having a "voice", which in turn is the result of having attributes such as power, legitimacy and urgency in relation to an issue.[40]

Ramírez (1999) has developed a conceptual framework for stakeholder analysis and conflict management (see Figure 5) based on a number of propositions and is intended to guide enquiry. The propositions are derived from an extensive review of the literature and reference to the original text of Ramírez is recommended. The framework includes both stakeholder considerations in conflict and non-conflict situations.

Figure 5. Conceptual framework for stakeholder analysis and conflict management (from Ramírez, 1999)

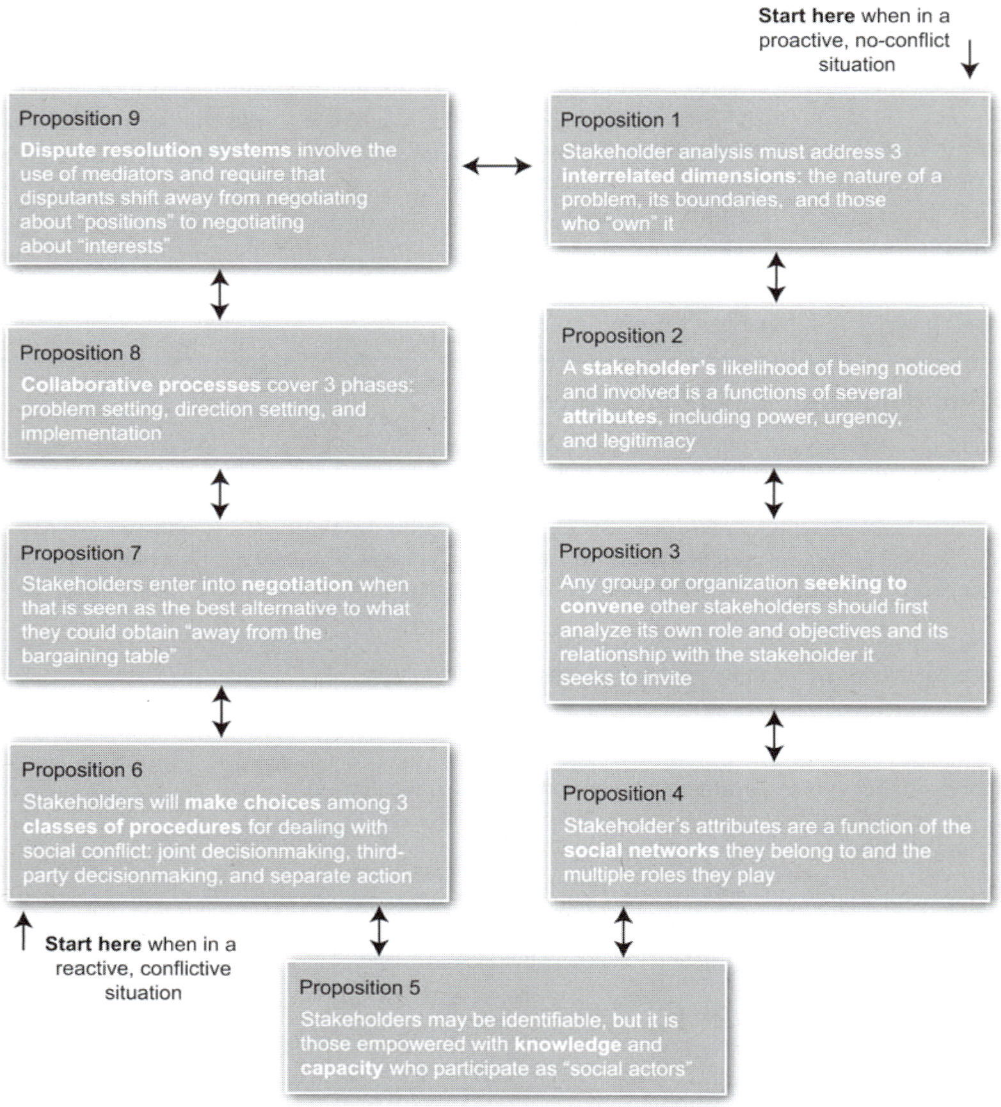

[40] Ramírez, 1999.

Box 7. Stakeholder identification and analysis (cont.)

Stakeholder analysis tools themselves tend to be straightforward lists or tables of criteria or dimensions against which groups occurring in the area are assessed. While it is relatively straightforward to generate these tables, more complex and constantly changing are establishing an agreed definition of issues or problems, defining the boundaries of the problems as well as identifying which groups are relevant to the situation.

In assessing forest stakeholders, Colfer (1995) identified some of the most important elements or dimensions in people-forest interactions (below), adapted for wider relevance to sacred natural sites. These dimensions are:

- **Proximity (physical and emotional)**
This dimension refers in the first instance to physical closeness to the resources, and weights stakeholders who live nearby. However, many people who have a stake in sacred natural sites may for a variety of reasons, live at a distance. This is particularly true when broader sacred geographies are taken into account.

- **Pre-existing rights**
The recognition of pre-existing rights is important from ethical and pragmatic viewpoints. Ignoring these rights often results in contested and conflict situations.

- **Dependency**
Many communities are heavily dependent on the resource in question, for spiritual or sacred values. This dependency may be less on direct resource use but related to the cultural values and underpinnings of the society in question, increasingly recognised as an ecosystem service.

- **Indigenous knowledge**
Many key stakeholders have extensive knowledge about particular areas, usually as received knowledge from earlier generations and/or experience borne of long association. The demonstration of local knowledge about a site also establishes the links with specific communities.

- **Resource/culture integration**
Cultures tend to be intimately linked to their environment. Sacred natural sites are one of the strongest manifestations of this and often give meaning to life and a sense of self to individual members of a community.

- **Power deficits**
In many cases groups that are associated with a particular resource may have relatively little power, especially if the resource has been incorporated within a legally designated protected area. Recognising and addressing power deficits is an important part of stakeholder analysis.

A careful understanding of those institutions or individuals with a stake in a particular area is very important, in order to ensure that all key stakeholders are included in the negotiations about, and the management of, sacred natural sites.

3.4 Legitimacy

Recognise that different groups have different levels of legitimacy in decision making about sacred natural sites.

Of the multiple stakeholders involved in sacred natural sites, some will have a greater and more legitimate voice than others in discussions and decisions. Of particular importance will be appropriate traditional custodians. Ensuring that those with the greatest stake in management decisions are heard is an important element. A practical approach is to assign stakeholders to primary, secondary and tertiary groupings as recommended by the ecosystem approach (guideline 2.7).[41]

The complexities of custodianship need to be well understood. Custodianship of some sacred natural sites may be numerous and multifaceted, and a careful, in-depth and respectful understanding of the custodians of a site, as well as of the broader community context, is necessary before embarking on an appropriate process of engagement. It is important to identify the authentic custodians and authorized decision makers. This will require careful work and relationship building over time to fully understand community dynamics, especially where sites are important to different communities in which multiple cultures and custodians are involved.

[41] Shepherd, 2003; 2004.

3.5 Conflict management

Where relevant and appropriate, use conflict management, mediation and resolution methods to promote mutual understanding between traditional custodians and more recent occupants, resource users and managers.

While in many cases different stakeholders come to mutual accommodation over the use of sacred natural sites,[42] conflict does occur and sites do become contested. Situations arise, for example, where there is competing authority over sites. It may be necessary to use mediation and conflict management techniques to come to mutual understanding. While it is hoped that through these efforts there will be improvements in the situation, it may be unrealistic to expect that the problem will be fully solved. A proportion of stakeholders may be unsatisfied with the outcome. In some situations, protected area agencies may be able to play a role in conflict management as mediators. In other instances, protected area managers may be too closely involved and a third party mediator may be required. Careful assessment will be needed to ensure that protected area authorities do not exacerbate sensitive situations. Support NGOs have often played valuable roles in these circumstances. In the case of managing conflict over climbing at Devils Tower National Monument in the United States (Case study 6), conflict was managed in the first instance through several initial stakeholder meetings, and then in the court of law. Following a legal decision, more positive relationship building began between different contesting groups.

Governance, mediation and conflict management tools are widely available and are evolving (see Box 7 for a framework on stakeholders and conflict). Negotiation methods can be very helpful, as can recommendations from experienced and trained mediators (see Annex 8). In addition, public education can play a critical role in minimizing or avoiding conflict (see principle 4).

Case study 6. Devils Tower National Monument (Mato Tipila), USA

Devils Tower National Monument, America's first national monument, rises above the surrounding grassland and ponderosa pine forest and can be seen from miles away. Located on the western edge of the Black Hills, the Tower is not only a spectacular geological wonder, protecting mountain and Northern Plains animal and plant communities distinctive to the Black Hills, but is also of major cultural significance. Known by Northern Plains Tribes as *Mato Tipila* – Bear's Lodge – the Tower is a sacred site of worship for many American Indians. It is also the beacon that welcomes home local residents. In addition, climbers have long acclaimed Devils Tower as one of the premier technical climbing sites in North America.

In 1995, a climbing management plan for Devils Tower was produced in an effort to address the interests and needs of all those using the Tower, as well as the impacts from those uses on natural and cultural resources. The plan grew out of a series of meetings between park staff, climbers, environmental groups and Native Americans, which resulted in the implementation of a voluntary closure to climbing during the month of June. Further, federal laws and Executive Orders allow American Indians the right to practise their traditional religious and cultural activities on federal lands.

The 1995 Climbing Management Plan (CMP) attempts to balance recreational use by climbers and traditional cultural use by American Indians. In respect for the reverence that American Indians hold for the Tower, the National Park Service (NPS) established and advocated the voluntary climbing closure for Devils Tower during the month of June, when many traditional cultural activities occur (i.e., summer solstice ceremonies). The voluntary closure was primarily established to promote greater understanding of American Indian tribes culturally affiliated with the Tower as a significant Sacred Site.

The 1995 climbing management plan sparked a 1996 lawsuit, *Bear Lodge Multiple Use Association vs. Bruce Babbitt*. The issue of the Tower as a Sacred Site for American Indians was central to the court case, in which the plaintiffs alleged that the June voluntary closure promoted religion and violated their Constitutional 1st amendment rights. The Wyoming federal district court rejected the plaintiffs' claims in April 1998, holding that the plan was a legitimate exercise of agency discretion.

The NPS reviewed the 1995 climbing management plan in 2002. The CMP update was finalized and released to the public in 2006 with the following features:

Climber education programme: An improved climber education programme will continue to include climbing ranger(s) on the staff and a dedicated climbing registration office with a climber education video. The climber education video will be available to increase climber awareness of safety, resource impacts and the Tower's cultural significance.

[42] Dudley *et al.*, 2005.

Safety standards: The NPS will require professional certification for commercial guides that trains and tests guides for competency. This standard will improve safety for guided climbers which make up about 40% of Tower climbers.

June voluntary climbing closure: The NPS will continue a voluntary climbing closure on the Tower during the month of June when many traditional cultural activities occur. This voluntary closure has been in effect each June since the CMP was established in 1995.

Other elements of the 1995 CMP remain, including: a seasonal climbing closure to prevent disturbance to nesting falcons; and allowing climbing bolts to be replaced using hammers. New bolts and the use of power drills are not allowed.

In spring 2006, the centennial of the establishment of Devils Tower National Monument, the first American Indian was hired as Superintendent. This has led to greater input and participation by local community members, climbers and tribal members on management decisions within the Park. Throughout all the years of conflict, large inclusive meetings were never held among the tribes, climbers and local community members. This has now changed, and it is anticipated that meetings will be held with mediators to facilitate discussions with all users involved in issues related to climbing activities. Within the past year (2007), several significant events have occurred that brought together the different parties, as a prelude to such meetings. A shared meal was held at a local climbing guide's home, with Chief Arvol Looking Horse and local community members in attendance. During the month of June, a local medicine man invited Park Service employees and a climbing guide to participate in a sweat lodge ceremony with tribal members. This has initiated more constructive relationships and new cooperation among the affiliated tribes, local communities, neighbours and the climbing community.

This success has been accomplished in part due to the recognition of a common foundation amongst users for the protection of the Tower for future generations. It is our desire that this foundation be built upon, resulting in a respect for one another that will transcend past differences and bring users together through partnerships and collaborative efforts to provide greater protection for the Tower and its natural and cultural resources.

Author: D. FireCloud, Superintendent, Devils Tower National Monument, US National Park Service.

4 Encourage improved knowledge and understanding of sacred natural sites.

4.1 Multidisciplinary approach

Promote a multidisciplinary and integrated approach to the management of sacred natural sites calling on, for example, local elders, religious and spiritual leaders, local communities, protected area managers, natural and social scientists, artists, non-governmental organizations, and the private sector.

Sacred natural sites require an integrated research and management system. Since sacred natural sites combine the preservation of the environment and its biotic resources, as well as the living cultural manifestations and belief systems of local and indigenous communities, a truly integrated management system is needed. This management system will ideally be capable of understanding, and caring for, both the natural and the cultural space. In this vein, a holistic management scheme should be put into place that satisfies conservation aims, cultural conservation objectives, and expression of spiritual belief systems (see Box 8). Ecologists, cultural anthropologists and traditional practitioners can combine their efforts to ensure integrated management of the natural environment and the socio-cultural dimensions of the area based on a multidisciplinary and cross-cultural understanding. In these situations there may be a need to draw upon the arts and humanities – linguists, philosophers, religious and spiritual leaders, poets, writers and artists – to fully understand the spiritual dimensions of a sacred place. It is likely that artists, for example, will be in a good position to engage with the wider public to support public awareness and understanding of sacred natural sites.

At Hawai'i Volcanoes National Park, a World Heritage Site in the USA, a painting of Pele, goddess of the volcano, hung for many years in the Kilauea Visitor Center. Pele is volcanism in all its forms, and her present home is Kilauea and Mauna Loa within the park. These are among the most sacred of mountains in the world. Native Hawaiians felt that the painting, done in the early 20th century, did not reflect their culture, and that Pele looked like a Euro-American. In partnership with The Mountain Institute and the park's *kupuna* group (Hawaiian elders), a competition was held to create a new painting of Pele. The competition was intended to raise visitors' awareness of the sacredness of the volcanoes and their living deity Pele. The park was overwhelmed with the number of paintings received from local artists, a show was mounted, a winner was selected, and the old painting was replaced with a new representation of the creative force that spawned the protected area (see photo plates).

Box 8. Evaluating the cultural and spiritual values of a sacred natural site

Within each protected area, it is recommended that management plans include guidelines and procedures to assess the local importance of "intangible" cultural and spiritual values, such as:

- traditional, indigenous knowledge and local practices, including: medicine, botany, zoology, animal husbandry, agriculture, water stewardship, crafts, mapping, oral history;
- cultural heritage and identity, traditions related to place that are embedded or expressed in language, music, song, dance, art, literature, festivals, foods;
- spirituality, rituals, *cosmovision* (world vision or worldview);
- community values expressed in social institutions, such as:
  - customary law systems;
  - consensus decision making;
  - harmony with nature;
  - sustainable use of resources;
  - nonviolence, compassion, co-habitation, peaceful relations with people.

Protected area management policies should consider, and public education materials should include mention of, the following specific types of cultural and spiritual values ("intangible values"), and how they may be embodied in the protected area landscape:

- intrinsic, existence value
- health and healing value
- recreational values of tourism, wilderness experience
- educational value
- western scientific knowledge or history
- sense of place
- aesthetic quality, scenic values
- spiritual inspiration, pilgrimage
- peace and reconciliation
- artistic inspiration and expression

If a discrete sacred natural site, a wider area, or an extensive landscape is deemed important because of:

- biodiversity value due to high endemism, pristine water, sustenance, etc., geophysical formations or landscape values; and/or
- cultural value due to medicinal plants, sacred animals, ceremonial use, vision questing, pilgrimage, oral tradition, etc.; and/or
- spiritual value due to power, solitude, minimal use, etc.;

then protected area managers should develop conservation strategies for that site or area *based on the foundation of those specific biological, cultural and spiritual values.*

Author: C. McLeod.

4.2 Integrated research

Develop an integrated biological and social research programme that studies biodiversity values, assesses the contribution of sacred natural sites to biodiversity conservation, and understands the social dimension, especially how culturally rooted behaviour has conserved biodiversity.

Sacred natural sites contain significant biodiversity values and appreciation of the role of custodians in protecting nature is growing. While there is an increasing body of research, the actual extent of these biodiversity values, and the relationship between biodiversity and sacred natural sites, is still not well elucidated. Likewise, the contribution that sacred natural sites and their associated cultural practices make to protecting and reducing the loss of biodiversity is still not well understood. Studies that have shown the conservation values of sacred natural sites have helped to broaden the constituency of people who care about their survival. Research programmes that are respectful of custodians' wishes can increase knowledge in this area and help to scientifically establish the relationship between sacred sites and biodiversity. This knowledge will help promote the important role that sacred natural sites play in biodiversity protection and this, in turn, will enhance management policies and practices intended to protect sacred natural sites (see section 3.4).

4.3 Traditional knowledge

Consistent with article 8(j) of the Convention on Biological Diversity (CBD), support the respect, preservation, maintenance and use of the traditional knowledge, innovations and practices of indigenous and local communities specifically regarding sacred natural sites.

Under Article 8(j) of the CBD, Parties to the Convention undertake to respect, preserve and maintain the knowledge, innovations and practices of indigenous and local communities embodying traditional lifestyles relevant for the conservation and sustainable use of biological diversity. They also undertake to promote the wider application and encourage equitable sharing of resulting benefits of such knowledge, with the approval and involvement of its holders.

Both science and traditional knowledge should be fully utilized for the conservation and management of sacred natural sites. Integrated management schemes should call upon the use of natural and social sciences as well as the use of traditional knowledge. As regards traditional ecological knowledge, many custodians of sacred natural sites have a wealth of knowledge on the biophysical environment in their roles as protectors of sacred species, as traditional healers and herbalists, and as decision makers in the context of the agricultural calendar. While respecting and protecting the intellectual property of indigenous cultures, the sharing of science and traditional ecological knowledge should be beneficial for all stakeholders in the sustainable management and conservation of sites. Custodians of traditional knowledge, natural and social scientists, and students of the humanities should be encouraged to work together in an integrated manner to ensure the sustainable safeguarding of sacred natural sites.

Berkes (1999) considers traditional knowledge at four interrelated levels: 1) local knowledge of land and animals, 2) knowledge of land and resource management systems, 3) knowledge and analysis of social institutions, and 4) knowledge and analysis of world views (see Figure 6). These four levels provide a useful framework for considering traditional knowledge of particular relevance to sacred natural sites. Berkes and others[43] also recognise the important relationship between traditional knowledge and spiritual values:

> "The globalization of Western culture has meant, among other things, the globalization of Western resource management. The remaining pockets of traditional

Figure 6. Four levels of analysis in traditional knowledge and management systems (from Berkes, 1999)

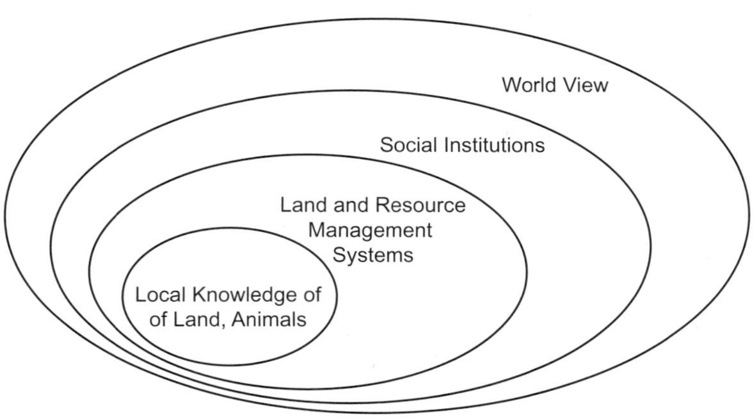

[43] E.g. Ramakrishnan, 2003; Khamaganova, 2007.

systems probably cannot escape history, but they can inspire new approaches to environmental stewardship and suggest more participatory and locally grounded alternatives to top-down, centralized resources management … Perhaps the most fundamental lesson of traditional ecological knowledge is that worldviews and beliefs do matter. Almost all traditional ecological knowledge systems may be characterised as a complex of knowledge, practice ***and belief***. Almost universally, one encounters an ethic of a nondominant, respectful human-nature relationship, a sacred ecology, as part of the belief component of traditional ecological knowledge".[44]

4.4 Networking

Facilitate the meeting of, and sharing of information between, traditional custodians of sacred natural sites, their supporters, protected area managers and more recent occupants and users.

To strengthen the management of sacred natural sites it is often beneficial to bring together custodians from different sites and even different traditions. This allows the sharing of ideas and experiences. Many traditional custodians of sacred natural sites operate in isolation. Linking custodians of different sites to share experiences supports their efforts to protect their sacred natural sites. Networking can take place within countries as well as at other geographical scales. In cases where landscapes containing sacred sites are now under the management of more recent occupiers and users, it may be useful to bring together different users to explore mutual understanding and improved management of cultural landscapes.

4.5 Communication and public awareness

Develop supportive communication, education and public awareness programmes and accommodate and integrate different ways of knowing, expression and appreciation in the development of policies and educational materials regarding the protection and management of sacred natural sites.

For many sacred natural sites, the level of protection they receive is in direct relation to the degree that they are recognised, understood and supported by the surrounding societies. To engage wider groups within those societies and to increase the understanding and support for sacred natural sites, dialogue should be initiated through communication, education and public awareness activities. Many communities have skills in traditional approaches to communication, especially those that still embrace oral traditions. Current thinking in the fields of communication and education emphasises "two-way learning". Drama, dance, song and poetry can all be effectively used as two-way communication, engaging community members themselves as well as other people. These performing arts are also a means of holding and expressing local intangible heritage, knowledge and ethics (see Case study 9, Malawi). Drama and theatre can be very effective at expressing sensitive issues in non-contentious ways, not least in informing protected area managers of their shortcomings. These approaches are also culturally rooted and can help keep local cultures and art forms vibrant. Other approaches include museum displays, travelling exhibits and the use of festivals, which have all been used in India to promote knowledge about sacred groves (Case study 7, Indian sacred groves). The Indian National Museum of Mankind has also planted nine replica sacred groves as an innovative living demonstration.

Within IUCN, the Commission on Education and Communication (CEC) provides a wide range of materials of relevance to protected area managers. Increasingly, information about the sacred values of protected areas is being included in protected area interpretation programmes (Case study 15, Dhimurru IPA).

Communications, education and public awareness (CEPA) is a cross-cutting theme under the CBD, which has a programme of work and priority actions being developed toward the 2010 international biodiversity target. The CBD operates a CEPA "portal" to provide information on public awareness.[45]

4.6 Inventories

Subject to the free, prior and informed consent of custodians, especially of vulnerable sites and consistent with the need for secrecy in specific cases, carry out regional, national and international inventories of sacred natural sites and support the inclusion of relevant information in the UN World Database on Protected Areas. Develop mechanisms for safeguarding information intended for limited distribution.

If local custodians deem it wise, an inventory of sacred sites, confidential if appropriate, could be created to guide management and conservation policies. This might include the participatory mapping of sacred natural sites, as well as recording details about their use and historic management (guideline 1.4). In certain instances it would be useful to know the total number of sacred natural sites in a region. This would help evaluate their contribution to nature conservation as well as to cultural values, and would

[44] Berkes, 1999, p.163, original emphasis.
[45] CEPA portal: http://www.cbd.int/programmes/outreach/cepa/home.shtml

Case study 7. Communicating the importance of Indian sacred groves to a broader audience

Realizing the cultural, biological and ecological importance of the sacred groves in our country and the threats faced by this ancient institution, the National Museum of Mankind (Indira Gandhi Rashtriya Manav Sangrahalaya – IGRMS), Bhopal City in central India has undertaken a number of activities in collaboration with several other institutions:[46]

Nine living replica sacred groves for education: In 1999, the museum installed at its 200-acre campus replicas of sacred groves from the states of Arunachal Pradesh, Chhattisgarh, Kerala, Maharashtra, Manipur, Meghalaya, Rajasthan, Tamil Nadu and West Bengal. Plants from groves of these states were planted in their respective replicas, after careful selection, taking into account local soils and climate. These groves were ritually established, accompanied by dances and ceremonies performed by the local communities of the respective states. One grove was established by the residents of Bhopal City following the community tradition. These replicas serve as living nurseries of ancestral and community identity, purity, longevity and sacred values.

Indoor exhibit: An indoor exhibition has been developed on sacred groves, using photographs, maps and charts depicting various aspects of Indian sacred groves. It is hoped that visitors to the museum will learn about the sacred grove tradition.

Travelling exhibit: A travelling exhibition was created in 1999–2000, using 67 panels of photographs and maps.[47] The objective of this travelling exhibition was to interact with local people and different organizations to learn more about the sacred groves of the country and to strengthen the diverse local management practices and knowledge systems. The exhibition has elicited an excellent response and is being enriched with further information from the different places. It is in demand all over India.

Sacred Grove Festival: A three-day National Sacred Grove Festival was organized in January 2000, involving 185 participants from 15 states. The festival provided, for the first time, a platform for different stakeholders, such as grassroots level functionaries associated with sacred groves including foresters, scientists and media people to discuss various aspects of the sites, including the formation of a network of stakeholders, developing region-specific field-based activities and the publication of relevant information.

A sacred grove festival was again organized in March 2007 to revive the network and a book entitled *Sacred Groves in India* was released on this occasion. Such festivals provide an opportunity for the stakeholders of the sacred groves to come together and share experiences of managing and conserving sacred groves in their respective regions. As a follow up to this, IGRMS has decided to develop the initiative on sacred groves further as part of the activities of the museum.

Demonstration district planning: As a follow up to the national-level programme, a model example of developing and implementing a sacred grove conservation strategy was demonstrated in Kodagu district of Karnataka. In Kodagu, *devarakadus*, as sacred groves are known, have had the status of Protected Forests since the Indian Forest Act of 1878. Despite this, *devarakadus* have, over the years, been damaged by illicit cutting, planting with exotic species, grazing, fire and encroachment. As a response, the State Forest Department decided to join forces with local people, the managers of sacred groves, to improve the protection of the sacred groves in Kodagu district. The State Forest Department undertook a consultative process with the active involvement of the Forestry College at Ponnampet in Kadogu district, which had earlier undertaken extensive research on various aspects of *devarakadus*. Through deliberations of working groups of sacred grove managers constituted for the purpose, a proposal evolved known as "Joint Forest Protection and Management – Devarakadu". The objective of the proposal was to involve villagers in the planning, protection, regeneration, development and management of the Devarakadus Protected Forest areas. The initiative was launched on the occasion of the Devarakadu festival at Virajpet, Kodagu in October 2000. Villagers who are traditionally bound to these forests have constituted themselves into sacred grove committees. The formation, composition, duties and responsibilities, the produce-sharing mechanism and other modalities of sacred grove committees, and the umbrella "Federation of sacred grove committees of Kodagu District" were established. At present, the programme is recognised by the State Forest Department but further awareness raising among sacred grove committees is needed to improve on the programme's success.

Author: Yogesh Gokhale, The Energy and Resources Institute, New Delhi, India.

[46] Including the Indian Statistical Institute, Calcutta; Centre for Interdisciplinary Studies, Barrackpore; Dept of Anthropology, University of Pune; Indian Institute of Science, Bangalore; Kerala Institute for Research and Training in Anthropology and Development Studies, Kozhikode; St Joseph's College, Thiruchirapalli; Applied Environmental Research Foundation, Pune; Indian Institute of Forest Management, Bhopal; North Eastern Hill University, Shillong; World Wide Fund for Nature-India.

[47] Malhotra *et al.* 2000.

assist in the development of appropriate policies at different levels of governance. A good example of a biological inventory leading to conservation is Case study 5 on Kenya's Kayas.

At the global level, the actual number and extent of sacred natural sites is unknown. In terms of number they certainly exceed the current number of protected areas. Many sacred natural sites are small, but some are quite extensive. Five of the planet's 20 largest protected areas are locally designated as indigenous (this does not imply, necessarily, that they are deemed "sacred").[48] Regional, national and global inventories of sacred natural sites will aid our understanding of them. Likewise, the rate of loss is not known and it is important to understand the patterns, the rate and the causes of loss of sacred natural sites in order to find solutions and support their conservation.

Listing sacred natural sites, considered for inclusion or located in protected areas, in the UN World Database on Protected Areas would add to the global understanding of these sites as they relate to legally recognised protected areas. If these sites are listed, care should be taken that the listing is with the free, prior and informed consent of local communities and their spiritual leaders. Proper safeguards would have to be put in place to effectively safeguard sensitive information for limited distribution. Strict maintenance of confidentiality will be critical to building trust and achieving successful collaboration.

4.7 Cultural renewal

Recognise the role of sacred natural sites in maintaining and revitalizing the tangible and intangible heritage of local cultures, their diverse cultural expressions and the environmental ethics of indigenous, local and mainstream spiritual traditions.

In some countries there has been an erosion of traditional management of sacred natural sites. Younger generations may be losing their connection with community traditions. Sacred natural sites can play a fundamental role in the revitalization of community values and traditions. Support from governments and civil society to custodians can help them maintain both the management of sites and the continuity of culture. In some situations efforts to rebuild cultural connections to nature are following periods of suppression of these cultures (Case study 8).

Particularly important in cultural renewal is the promotion of indigenous education systems, within which sacred natural sites can play an important role. The UN Declaration on the Rights of Indigenous Peoples identifies the right of indigenous peoples to establish and control their own education systems. It is important in many other contexts of local communities for education systems to effectively pass on traditional community knowledge and understanding related to culture, nature and sacred natural sites. Again, Erjen Khamaganova tells us:

> "In my native Buryat language, the process of education is called 'Khumuuzhlekhe', from the word 'Kkun' – human. It literally means helping a person to become 'human' which in my native understanding is the ability to comprehend the world in the entirety of complex interconnections and interrelations of events, phenomena and actions. Sacred sites in Buryat culture are the embodiment of these interrelations and are a major 'educational tool' to realize oneself as a human being, and as a part of an endless whole".[49]

Case study 8. Daoist Ecology Temple at Taibai Shan National Nature Reserve, Shannxi, China.

Taibai Shan Mountain, Shannxi Province, stands at the centre of the Qinling Range and covers a great expanse in the centre of the mountain range. It lies within the Taibai Shan National Nature Reserve, one of a growing network of protected areas set up specifically to conserve the habitat of the threatened Giant Panda, but which also protects a wide range of other animals and plants including the Ginkgo Tree, rare pheasants, salamanders and monkeys. Other protected areas in the region include the Zhouzhi National Nature Reserve, the Laoxincheng Nature Reserve, and the Xi'an Heihe Forest Park.

The whole range of Qinling is considered sacred to many Daoists and Buddhists in China, with over a million visitors to the area each year coming both as an act of faith but also to enjoy the beautiful surroundings. When conservation efforts to help the Giant Panda were initiated in 2003, supported by WWF-China, it was initially imagined that the most appropriate conservation effort would be a community-based programme that would raise awareness and bring about behavioural change. This

[48] In 2003 there were approximately 102,000 protected areas. India alone has an estimated 150–200,000 sacred groves, albeit mostly small. Protected areas cover 18.8 million km2 (Chape *et al.*, 2003) while sacred natural sites cover an unknown area. There will, of course, be a considerable but unknown overlap between the two sets of areas.

[49] Khamaganova, 2007.

conservation plan, however, did not include any recognition of the Daoist or Buddhist outlook and practices that had helped to maintain the ecology of the mountains for centuries and which are still very much alive in many of the visitors.

Sacred mountains, in fact, have a very specific cultural and religious significance that in many respects defines the ecological understanding of them. This religious understanding in turn helps define the principles by which the sacred mountains are managed for ecological conservation. In the mid-1990s a survey of the five main sacred mountains in China by the Daoist Association of China, with ARC, showed that in those mountains where Daoist monks were in residence, the environment was better protected.

Daoism actually measures the wealth of a society not materially but by its ecological riches as expounded in the Daoist statement on Ecology: "Daoism has a unique sense of value in that it judges affluence by the number of different species. If all things in the universe grow well, then a society is a community of affluence. If not, this kingdom is on the decline".[50]

Daoist-led conservation activities at Taibai Shan have firstly placed the recognition of the sacred values of the Qinling Mountains at the heart of conservation action, secondly have rebuilt a Daoist Temple, destroyed during the cultural revolution, as a training centre in ecology, and thirdly used these activities to expand this initiative to other sacred mountains along the Qinling Range.

The second stage of this project was successfully completed in June 2007 when China's first religious ecological training centre was opened.[51] The Taibaishan Tiejia Daoist Ecology Temple has been built on the site of a temple considered important to Daoists for a thousand years and which still receives visitors. It is near the site of a sacred tree, and close to a WWF visitor centre.

The opening of the Daoist Ecology Temple was followed by a second workshop which explored Daoism and conservation. Participants included Daoist monks and nuns (representing 18 temples), park staff and representatives of conservation organizations. At this workshop the first step to expand this effort beyond Taibai Shan was taken, when a new network, Daoist Temple Alliance on Ecology Education was formed. Proposed activities of the Alliance include bringing the land use and management of the temple into the ecology and protection of surrounding areas, and developing a Daoist Ecology Education handbook.

Source: Dudley *et al.*, 2005; ARC website: http://www.arcworld.org/projects.asp?projectID=257; and Xiaoxin, 2007.

4.8 Intercultural dialogue

Promote intercultural dialogue through the medium of sacred natural sites in efforts to build mutual understanding, respect, tolerance, reconciliation and peace.

Sacred natural sites present an opportunity to develop mutual understanding between different groups.

Certain examples show that custodians of different cultures have been able to accommodate each other in a peaceful and collaborative way over long periods (Case study 11, Sri Pada-Adams Peak). Efforts need to be made to uphold this principle. Protected area agencies may be able to support and enhance this collective collaboration. They can serve to build bridges for intercultural dialogue and understanding while promoting tolerance, cooperation, reconciliation and peace. Examples exist, however, where protected area agencies or employees have exacerbated existing tensions and this, obviously, should be avoided.

Elements of intercultural dialogue through sacred natural sites can include:

- Mutual learning and "both ways" activities (Case study 15, Dhimurru IPA);
- Shared events and festivals;
- Joint research and exploring different ways of knowing;
- The increasing recognition and use of older or original place names, to restore their significance and meaning;

Many sacred natural sites have been contested and are the focus of conflict. Intercultural dialogue, understanding, the promotion of tolerance, cooperation, reconciliation and peace building are all elements of reducing conflict (guideline 3.5).

50 China Daoist Association, Beijing, China in Palmer and Finlay, 2003.
51 Xiaoxin, 2007.

Case study 9. Global intangible heritage and the Chewa sacred groves, Malawi

In 2005, the *Gule Wamkulu*, the Great Dance of Malawi, Zambia and Mozambique was proclaimed by UNESCO as a Masterpiece of Oral and Intangible Heritage of Humanity. Gule Wamkulu is intimately linked with the Chewa sacred groves and provides a direct link between intangible heritage and sacred natural sites.

"Gule Wamkulu is both a secret cult and ritual dance practised among the Chewa people living in Malawi, Zambia and Mozambique. It is performed by members of the *Nyau* brotherhood who are responsible for the initiation of young men into adulthood, and the performance of the Gule Wamkulu at the end of the initiation procedure to celebrate the young men's integration into adult society.

The Nyau dancers wear costumes and masks made of wood and straw, representing a great variety of characters, such as wild animals, spirits of the dead, slave traders as well as more recent figures such as the *honda* or the *helicopter*. Each of these figures plays a particular, often evil, character representing certain forms of misbehaviour in order to teach moral and social values to the audience. Nowadays, even if the Chewa matrilineal system has lost its social significance, Nyau societies, and with it Gule Wamkulu, are still very much alive and Chewa men tend to be both members of a Christian church and a Nyau society".[52]

The Chewa sacred groves are small areas of natural and undisturbed vegetation, isolated amongst areas of intensive agriculture and other modified land use. They support indigenous trees that often attain considerable size, and in many places are some of the few remaining refuges of forest biodiversity. Known locally as "Graveyard Forests" they have two parts: a graveyard and mask yard. Community members are buried in the graveyard, and funerals are the only time when all members of the community are allowed into the sacred groves. On other occasions it is only initiated male members of the Nyau society who are allowed to enter. In the mask yards, society members make masks and practise the Gule Wamkulu dances. Dancers and masks emerge for key events (see photo plates).

Increasingly the traditional leadership and Gule Wamkulu are working with ministries, for example with the Department of Health, on the promotion of primary health care messages. A similar initiative is under consideration by the Forest Department as part of its participatory forestry programme. Most forest reserves, especially in the south of the country, have been badly degraded in contrast to the Graveyard Forests, although these are now under severe pressure themselves. This is due to the fact that they provide wood for coffins, the demand for which has dramatically increased due to the AIDS pandemic. In response to the pandemic, the role of the traditional leadership is now crucial to changing sexual attitudes and practices.[53] A return to the use of traditional reed coffins might be one solution to alleviate some of the pressure on the sacred groves. More generally, in Malawi's critical chronic state of food insecurity, environmental degradation and the HIV/AIDS crisis, the country may find one part of the solution is to explore synergies with strongly rooted traditions.

The understanding of the role of these traditions, including the Gule Wamkulu, has been greatly assisted by the Mua Mission of the White Fathers, and supported by the inculturation policy of the Catholic Church.[54] At the Mua Mission, Dedza District Malawi, the Ngoni Arts Centre has promoted traditional arts and crafts, and its museum has a collection of over 400 traditional masks. It has an active dance troupe that performs traditional dances and runs courses on cultural awareness for development workers, as well as other health and education projects. It supports outstanding woodcarvers who use both traditional and Christian motifs, some of whose works are now in collections across the globe, including the Vatican Museum. The work of the Mua Mission has greatly assisted the mutual understanding between Traditional African Religion and Christianity in the area, and supported the UNESCO proclamation of Gule Wamkulu as intangible global cultural heritage.

Author: R. Wild from interviews with Mua Village Head man and Nyau society members/gule dancers and the Director of the White Fathers Mission, Mua, Father Claude Boucher Chisale.

---

52 Adapted from UNESCO http://www.unesco.org/culture/intangible-heritage/22afr_uk.htm
53 Bryceson and Fonseca, 2005.
54 Ott, 2000; Boucher, 2002a and b.

5 Protect sacred natural sites while providing appropriate management access and use.

5.1 Access and use

Develop appropriate policies and practices that respect traditional custodian access and use, where sacred natural sites fall within formal protected areas.

Policies and practices are needed to support the access of traditional custodians to sacred natural sites when these sites have been incorporated into legally protected areas. Of particular importance is access to sites for performance of ceremonies and maintenance of individual sites. In some cases, the custodians of sacred natural sites that have been incorporated into protected areas have had to enter unseen or illegally to avoid protected area patrols while visiting their sites. In other cases, custodians are expected to follow complex and often unrealistic permitting procedures, or to pay entrance fees to access their sites.

In many protected areas that charge for visitor entrance, local community members enter for free or at significantly reduced rates. If not already considered, this option should be taken into account when custodians and practitioners seek to visit and use their sacred natural sites.

Many spiritual practices associated with sacred natural sites require privacy, silence and serenity. Disturbance of practitioners can disrupt important ceremonies or years of spiritual endeavour. Due to increasing recreational pressures on protected areas and efforts to increase public access to nature, friction may arise between the privacy needs of custodians and the perceived freedoms of other visitors. This can be exacerbated where there is limited respect for the culture of the custodians. Therefore, policies should be enacted that allow for the periodic closure of areas within and around sacred natural sites to allow for privacy in the conduct of ceremonies at the site.

Case study 10. Indigenous efforts to recover sacred sites and strengthen indigenous culture at the Ka'sankwa, the Sierra Nevada de Santa Marta, Colombia

The indigenous people of Ka'sankwa, the Sierra Nevada de Santa Marta, are taking steps to recover, rehabilitate and restore both environmentally, culturally and spiritually four sacred sites located in the intensively settled coastal and lowlands of the Sierra. Facing considerable challenges that include the activities of guerrillas and paramilitary groups, large-scale banana and oil palm growing, deforestation, land-use change, large development projects, and now retreating glaciers and drying rivers, their objective is to restore the ecological and spiritual balance of the overall socio-ecological system, to strengthen their cultural tradition and to facilitate the ongoing spiritual and social work of the Mamos, the indigenous spiritual leadership.

Ka'sankwa, the Sierra Nevada de Santa Marta, is a mountainous massif and the highest coastal peak in the world. It is 5,775m (18,942ft) above sea level and has an area of over 12,000km^2. Ka'sankwa's high snow-clad peaks and glaciers feed water to 29 rivers, the agricultural productive lowlands and their 1,000,000 inhabitants, including the cities of Santa Marta, Riohacha and Valledupar, and eventually the Caribbean Sea. The higher altitudes of Ka'sankwa are within two National Parks (Sierra Nevada de Santa Marta and Tayrona), and three indigenous reserves. The whole area is recognised as a UNESCO Biosphere Reserve. The Sierra Nevada is one of the most distinctive, diverse and threatened areas in South America. Tapirs, Red Brocket deer and howler monkeys are among the 120 species of mammals along with elusive cats such as the Jaguar, Puma and Little spotted cat. The park also harbours 46 species of amphibian and reptile; those that live above 3,000m (9,900ft) are found nowhere else on the planet, having evolved in complete isolation. Six hundred and twenty-eight bird species have been recorded from the Sierra Nevada de Santa Marta National Park alone.

The 35,000 indigenous peoples, the Koguis, the Arhwacs, the Wiwas and the Kankuamos, are all descendants of the Tayrona civilization. They are heirs and custodians of the Tayrona traditions, and have managed Ka'sankwa, their ancestral lands, through cultural and spiritual "works", based on the Law of the Mother. These "works", including rituals and offerings, are conducted by the Mamos priest class at a network of sacred areas. These sites, located across the Tayrona traditional lands, are ranked according to importance. Some lie at upper altitudes, some in the mid-range and others in the lowlands and coast. According to indigenous accounts of the Earth's creation, the area around the Sierra was a circular territory with high mountains at its centre and a border, called the "black line", extending to the ocean where the water cycle ends. This territory is the centre of the world and home to the Mother's children who live off her and care for her water sources, lands and sacred sites. Some of the sites have *Cansamarías*, which are thinking, teaching and ceremonial houses in which the Mamos perform these and other works to ensure the maintenance of the territory and its culture, and to ensure the flow of spiritual forces between those sites and the centre of the Sierra and hence to maintain overall equilibrium.

Case study 10. Indigenous efforts to recover sacred sites and strengthen indigenous culture at the Ka'sankwa, the Sierra Nevada de Santa Marta, Colombia. (cont.)

From the first moment of contact with the Western world, the indigenous communities witnessed the incessant pillage and destruction of their territories, their sacred sites, burial grounds, and the customs of their ancestors. The four tribes that have managed to survive are undergoing various degrees of acculturation. Today, however, a small but growing number of people outside the indigenous community are starting to understand and value the exceptionally rich store of philosophical, ecological and spiritual knowledge that the Tayrona Mamos and indigenous community hold. This knowledge is not just of value in the wider Ka'sankwa catchment, but for humanity as a whole. The fact that effort is now being invested in understanding the basis for indigenous natural resource management indicates that the negative attitudes commonly held about indigenous knowledge during the colonial era have begun to change.

Over the years, the Mamos have endeavoured to maintain their spiritual practices at the network of sacred sites across Ka'sankwa. Today, however, non of the mid and low altitude sacred sites are under the stewardship of indigenous people. Many of these sites have been damaged, desecrated, are being used for other activities and are under other ownership. For several years, the growing dispute over territory between the guerrillas and paramilitary groups in the lower and middle parts of the Sierra, has not only severely limited community livelihoods but profoundly disrupted the practices and the limited management of sacred sites that the Mamos have been able to undertake.

The Mamos now feel it is appropriate to seek to recover four sacred sites, one for each community, and construct two Cansamarías in each (one for men and one for women) so that they, as guardians of memory and knowers of the spiritual and social laws and relationship to the Mother, can continue educating and transmitting the tradition of which they are heirs and protectors. In this way they also seek to reach out to the non-indigenous community, to build mutual trust and understanding and share their profound understanding of the Ka'sankwa, the Sierra Nevada de Santa Marta.

The first step in the process that the indigenous community wants to undertake is to survey, culturally and ecologically, the sites they no longer have access to, make links, connections and develop an understanding of and with the current owners. This would be with a view to exploring the possibilities for re-establishing four sites as functioning sacred areas.

The re-establishing of the four lowland sacred sites will support the physical and spiritual survival of the indigenous people of the Sierra, based on the constancy and continuity of the work that the Mamos perform, and is seen to be of vital importance. It will not only guarantee the permanence of indigenous culture in the ancestral territory, but also guarantees the protection of water resources and biodiversity on which the whole community of the Sierra depends.

Author: Guillermo E. Rodriguez-Navarro.

5.2 Visitor pressures

Understand and manage visitor pressures and develop appropriate policies, rules, codes of conduct, facilities and practices for visitor access to sacred sites, making special provisions for pressures brought about by pilgrimages and other seasonal variations in usage.

Many sacred natural sites are subject to high levels of visitor pressure. This can stem from their sacred, landscape, scenic and recreational value.

For some sites, visitor pressure may be constant throughout the year, but need different management during limited periods, for example, during festivals or pilgrimages (see Case study 11). Pilgrimages focusing on the site's sacred values can often be accommodated but can also present major challenges, especially if the nature of the pilgrimages change, the total number of pilgrims increases, or activities incompatible with the values of the site become widespread.

In some cases, the custodians of the sacred site live far from it. This requires considerable management support. An example of this is the Wirikuta Sacred Natural Site in the Chihuahuan Desert, Mexico, where the traditional pilgrimage of the Huichol people originates 135km away from Wirikuta.[55]

By the same token, some sacred natural sites demand restricted access and little or no visitation. Mechanisms should be put in place to control visitation and access in accordance with the wishes of custodians. For some custodians, ensuring that the site is "rested" is important (Case study 5, Kayas).

[55] Otegui, 2003.

Case study 11. Multi-faith accommodation at Sri Pada-Adams Peak, Sri Lanka

Sri Pada or Adams Peak is an area of high biodiversity and wilderness values and the catchment for several rivers that supply water to millions of people. It is the anchor of Sri Lanka's cultural landscape and a World Heritage Site. It is sacred to Buddhist, Christian, Islamic and Hindu communities. Several thousand people live on the fringes of its wilderness forest, and many of them are dependent on its resources. Meanwhile, three million people ascend the mountains each year to pay homage to what many consider to be a living being known by the royal title "Excellency". The focus of the mountain's sanctity is Sri Pada, a sacred footprint etched into the mountain's rock summit. Each community has its own beliefs regarding Sri Pada. For Buddhists it is believed to be the footprint of Lord Buddha from his third visit to Sri Lanka (*ca.* 577BC). He visited the god Sumana Saman, the mountain's protective deity, who took the sacred relic into his care. Hindus believe it is the mark of the Lord Shiva, left after his world-creative dance. Muslims believe that the relic is the footprint of Adam who, after he was cast out of paradise as a penance, stood on this spot for one thousand years, on one foot. Christians believe it is the footprint of Saint Thomas, who is said to have brought Christianity to Sri Lanka.

The conical Sri Pada peak rises to an elevation of 2,243m (7,360ft) and is the most outstanding feature of the island's central massif. Its commanding position and the sacred footprint enhance the profound spirituality of the peak. Surrounding the summit is a forest wilderness area (224km^2) known as Samanala-Adaviya and it is this that is seen as a living "Excellency" whose generous nature sustains human life, regulates climate, and replenishes soil and water. The water flowing from the mountain in streams and rivers is considered purified "holy water".

A network of Buddhist temples exists on the edge, and provides religious services to pilgrims, while monks occupy caves in the forest for meditation. Communities at the forest perimeter, who also have links to the temples, use the forest for food, house construction, fibre and utensils. Over 640 medicinal plants have been recorded. Respect for the mountain permeates life, and local people honour it before beginning their daily activities. Rules are enforced through practices with strong social grounding. Permission is sought before entering the sacred area on pilgrimage or to gather forest produce. There is an understanding that any destruction, over-exploitation or wasteful extraction may disturb the authority of Sumana Saman and violators risk getting lost, injury and even death. Every resource is used with respect, seasons for harvest and pilgrimage are clearly defined, and sensitive areas are noted as inaccessible. Resource extraction is guided by in-depth traditional knowledge and performed by the most knowledgeable and experienced members of the community. Younger generations are instructed to follow their elders to gain ancestral experience and knowledge.

One local resident who spent more than 56 years of his life living in a rock cave in the forest, surviving on only forest produce, has a holistic vision of his forest:

"Siripa Adaviya is superior to all manipulated systems; full of great powers. This should be treated subjectively. This is a huge living organism. It has produced various habitats for thousands of flora and fauna. This Excellency nurtures the life forms and maintains diversity across the terrain. It provides food, water and shelter and regenerates materials season after season and sustains conditions to support all living beings. Its superiority cannot be explained, but it should be understood. The sacred mountain forest has sustained itself through natural process and every plant and animal within it can be found to be useful".

Pilgrimages take place during December to May, but from tradition no one is allowed to climb during the remainder of the year. Unlike residents, pilgrims cannot take any material or harvest animals and plants. Pilgrims take a vow to follow the well established rules of pilgrimage, bathe in holy stream water prior to ascent, wear clean clothes, avoid eating meat and sing religious songs as they climb.

State ownership of Adams Peak, as for other forests in Sri Lanka, has been long accepted. Generally believed to be areas of great power, forests were donated by ancient kings to Buddhist settlements to be used for meditation. At Sri Pada these communities act as stewards of different areas of the forest wilderness, as well as establishing pilgrimage routes to the summit. Under the legal authority of the Department of Wildlife Conservation since 1940, potential conflict between legal ownership and traditional relationships has been tempered by the difficulties of separating the forest from local culture. Culture is not a peripheral issue but central to conservation. Culture has, however, been alienated from conservation measures and cultural values have only been appreciated in terms of specific isolated cultural sites. Key challenges for sustainability are accommodating the traditional custodians and their domain, as well as culture, as local instruments for effective management. Sri Lanka has mechanisms to conserve nature and wilderness but has yet to establish appropriate mechanisms to deal with the sanctity of forests in an integrated way.

Case study 11. Multi-faith accommodation at Sri Pada-Adams Peak, Sri Lanka (cont.)

Sri Pada-Adams Peak is a prime example of how traditional people have been able to safeguard a sacred mountain forest without recourse to legal institutions and policy interventions. This is in the context of widespread forest loss over much of the country in the last 130 years. It also exemplifies how different communities of different spiritual traditions can accommodate each other in the use and veneration of the same sacred natural site. Legends and beliefs have made it possible to embrace all groups under a common interest to safeguard and revere the sacred footprint in a collective and harmonious way.

Source: Adapted from Wickramsinghe, 2003, 2005.

5.3 Dialogue and respect

Encourage on-going dialogue among the relevant spiritual traditions, community leaders and recreational users to control inappropriate use of sacred natural sites through both protected area regulations and public education programmes that promote respect for diverse cultural values.

Many activities normally engaged in by visitors may be culturally inappropriate at sacred natural sites. Some examples of such activities are the climbing of sacred mountains or rock formations, entering into sacred caves or forests, bathing in sacred rivers, lakes or springs, participating in sacred ceremonies without the permission of the traditional practitioners, hunting of sacred animals, scattering of cremation ashes, leaving "New Age" offerings, or entering into sacred areas without permission or without culturally appropriate preparation. These activities can compromise sacred natural sites and disturb custodians and entire communities. In extreme cases they can lead to the custodians abandoning the site.

Various mechanisms can be developed to manage these activities. These include visitor education regarding acceptable behaviour, the use of zoning, establishing trails that steer visitors away from the most sensitive areas, and developing regulations that proscribe these activities (these are described in more detail in guidelines 2.4 and 2.5). Wherever possible, conflicts over use should be managed by dialogue between the different users so that compromises and mutual respect are achieved.

Education plays an important role in managing these conflicts and promoting dialogue and respect. In most cases, once visitors are aware of the behaviour expected of them, the great majority will respect the wishes of custodians by refraining from behaviours considered inappropriate or disrespectful to that particular sacred site. Controlling the persistent minority of people who do not respect custodians' wishes remains an ongoing challenge.

5.4 Tourism

Well managed, responsible tourism provides the potential for economic benefits to indigenous and local communities, but tourism activities must be culturally appropriate, respectful and guided by the value systems of custodian communities. Wherever possible, support tourism enterprises that are owned and operated by indigenous and local communities, provided they have a proven record of environmental and cultural sensitivity.

Tourism can provide an alternative source of income for communities, and with growing interest in traditional cultures, sacred natural sites can play a role in enhancing income and community livelihoods. Ecotourism, with its emphasis on supporting not only the resource but also the local communities, as well as promoting respectful and low impact visitation, is the ideal model. If practised well and managed with a guiding set of ethical principles, ecotourism linked to sacred natural sites can benefit local people directly – economically, socially and environmentally – but only if due respect is paid to indigenous and local peoples' value systems.

Tourism is now one of the largest economic activities on Earth, and demand to visit sacred sites and other "special places" is increasing every year. In many cases, sacred natural sites have to cope with major visitation. Monastic and other sacred sites of mainstream faiths in various countries receive large numbers of visitors. Likewise, renewed interest in sacred natural sites in China is increasing the visitation to these sites, sometimes with very large numbers of people. Tourism development projects have not always benefited or even involved the local custodians (guideline 5.5).

Learning, and sharing information, about tourism and sacred natural sites is needed by all stakeholders, especially the tourism industry. It is important, for example, that tourist brochures and information materials, as well as tour guides and leaders, accurately explain local cultural values to help enforce appropriate behaviours, restrictions and taboos. Examples might include whether or with what motivation to climb or refrain from climbing a sacred mountain, what to wear and how to behave when entering a sacred grove, and when not to take photographs of a ceremony or ritual.

If tourism activities are owned and operated by the custodians or their communities, whether they come from local, indigenous or mainstream traditions, there is a greater chance that tourism will be beneficial and support livelihoods. This can be positively reinforcing towards conservation both of the biological and cultural values of sacred natural sites. For example, studies on sacred groves in Ghana have shown that at the site where tourism benefits went to, and were managed by, the custodian communities, local people were much more supportive of the conservation of endangered monkeys living in the grove than the communities at a sacred grove where tourism management was run by the government and few benefits were received at the community level.[56]

5.5 Decision-making control

Strong efforts should be made to ensure that custodians of sacred natural sites retain decision-making control over tourist and other activities within such sites, and that checks and balances are instituted to reduce damaging economic and other pressures from protected area programmes.

Sacred natural sites are often very attractive to visitors. For a number of countries, sacred natural sites are international icons that draw millions of visitors each year. Unfortunately, tourism often undermines the very values that people come to experience. The double-edged nature of tourism is well known, yet effective management remains a challenge. The cultural, social, economic and environmental impacts of tourism range from the subtle to the profound. In the case of sacred natural sites, cultural sensitivities and potential risks are great, and therefore caution is needed. It is paramount that custodians retain control of sacred natural sites when decisions over tourism are being made. In a number of cases, the traditional custodians of such sites have had to struggle to regain a level of control to ensure that the integrity of their site is recognised and maintained.

The judicious application of the IUCN category system for protected areas can help in managing tourism. Category I protected areas are designed specifically for restricted or low-level access, and can be appropriate to support those sites where little or no access is required.

Once a site becomes of interest to tourists, additional rules will often need to be developed and applied to visitors. Recent regulations that have been generated in such situations include those related to behaviour, dress codes and photography. It is quite normal to restrict photography in specific locations, whether sacred or secular. Less common is restricting the type of photograph taken. Australia's Anangu people, traditional guardians of Uluru (Ayers Rock) prohibit the taking of photographs of many specific places within Uluru-Kata Tjuta National Park (Case study 12 and Annex 2). They request people take photos only of the whole rock, and park staff carefully evaluates all commercial proposals seeking to use images of the sacred site.

Zoning sacred natural sites can also help maintain control over tourism (guideline 2.5).

In some cases, custodians have actively sought to start tourism activities, as a means of supporting community livelihoods, and if carefully done this can be beneficial (see Case study 5, Kayas). Visitor interest in a sacred natural site and its associated culture(s) can help rekindle local pride in and increase the status of the sacred site.

Case study 12. Respecting culture at Uluru and Kata Tjuta, Australia

Australia's dramatic red monolith Uluru and its neighbour Kata Tjuta comprise an area of great spiritual significance to Anangu, the local Aboriginal people whose belief system is intertwined with the desert landscape. Renamed "Ayers Rock" and "Mount Olga" by European explorers and later promoted by the Australian government for conservation and tourism development, Uluru and Kata Tjuta were returned to Anangu ownership in 1985 and are protected within a 330,000-acre National Park (and World Heritage Site) and jointly managed with Parks Australia.

Many of the sacred places around Uluru are off-limits to tourists and photographers. While scores of tourists go to the rock specifically to climb it, an educational campaign discourages climbing, as stated on the park website: "Anangu have not closed the climb. They prefer that you – out of education and understanding – choose to respect their law and culture by not climbing."

The park's *Guidelines for Photography* state:

"Anangu believe that their culture is degraded if images of sacred sites are captured or displayed, or if the Park is used to advertise products and services that do not promote the natural and cultural values of the Park, and people's awareness, understanding and enjoyment of them ...

[56] Ormsby, 2007.

Case study 12. Respecting culture at Uluru and Kata Tjuta, Australia (cont.)

Uluṟu has great spiritual significance to Aṉangu. There are many sensitive sites associated with Uluṟu itself. We ask that you respect these places and their significance to Aṉangu by following these guidelines. Map 1 (see Annex 2) shows the parts of Uluṟu and specific sites that should not be filmed, photographed or painted, and locations from which images should not be captured. Most sites are clearly signposted, although some are not as they cannot be specifically identified or publicly discussed."

Aṉangu form a majority on the park's Board of Management and are employed as rangers and cultural interpreters. They have established management practices, based on their traditional cultural law, *tjukurpa*, to preserve the spiritual, cultural and ecological integrity of the park, and to ensure that their rights and knowledge are respected.

The restitution of the site to Aboriginal ownership, the reverting to the usage of Aboriginal names and the greater respect for the sacred Aṉangu values of Uluṟu and Kata Tjuṯa are significant steps forward, as are the sharing of some tourism revenue and the development of Aboriginal-owned tourist operations. Despite these advances, the social and economic conditions in the three Aboriginal communities around Uluṟu remain a major challenge.

(See Annex 2 for more detail)

Source: http://www.environment.gov.au/parks/uluru/vis-info/permits-image.html

5.6 Cultural use

While ensuring that use is sustainable, do not impose unnecessary controls on the careful harvest or use of culturally significant animals and plants from within sacred natural sites. Base decisions on joint resources assessments and consensus decision making.

The selective use of animal and plant species for ceremonial and other related purposes should be permissible in both a sacred natural site and the wider protected area, if harvests are sustainable and the overall quality of the environment is not jeopardised. Many traditional belief systems require the harvesting of plants or animals for a variety of reasons. These purposes can include regalia, offerings, festivals, sacred structures, as well as for subsistence during ceremonies. In some cases, the sought after animals and plants may be rare or endangered and these cases can be particularly challenging.

A case in point is the subsistence hunting in Dhimurru IPA, Northern Territory, Australia, of the endangered dugong and sea turtles, ocean-going mammals that the community considers sacred. In this case, the main threats to the dugong and turtles do not come from the community but from other groups. Dhimurru Land Management Corporation is developing sustainable harvesting protocols and promoting conservation of the animals amongst the wider community (Case study 15, Dhimurru IPA). In some cases, anomalies have arisen where, for example, local communities cannot access certain species for traditional regalia while the same species are being hunted by tourists in nearby recreational hunting enclaves, a situation that creates unnecessary resentment.

While traditions of harvesting plants and animals can be a source of conflict between protected area managers and local communities, efforts should be made to explore mechanisms by which the selective and sustainable use of such plant and animal species may be allowed. Increasingly, assessment tools using participatory techniques have been developed for decision making regarding this type of resource use.[57] As a principle, the use of species for commercial purposes should be prohibited, as this can rapidly lead to over-exploitation. In allowing resource use, however, attention should be given to avoiding the negative impacts that selective usage of biotic resources could have on the wider ecological system, and priority should be given to retaining the special qualities of the area.

If mechanisms for the selective use of culturally important species prove to be successful and sustainable, they can help build trust and confidence between protected area managers and local communities and may help to reduce conflicts over protected areas in general. The selective use of park resources has successfully been included in resource-use agreements between a number of protected area agencies and local communities.[58]

5.7 Protection

Enhance the protection of sacred natural sites by identifying, researching, managing and mitigating overuse, sources of pollution, natural disasters and the effects of climate change and other socially derived threats, such as vandalism and theft. Develop disaster management plans for unpredictable natural and human caused events.

Sacred natural sites face many challenges. These are unique to each situation but include threats stemming from changes in land use, increased demographic pressure, poverty and livelihood

[57] E.g. for plants by People and Plants International; see Annex 8.
[58] For example see Wild and Mutebi, 1996.

needs. There are also changes in value systems whereby younger generations grow less interested in local culture and religion. At the same time, there can also be periodic revivals in interest in traditions and the sacred sites associated with them. There are even situations in which elements of a local community are hostile to a sacred natural site for ethnic or religious reasons and vandalism can occur as a result. Occasionally, ceremonies are intentionally disrupted, sacred trees intentionally cut, sacred animals intentionally hunted, sacred mountains intentionally climbed, or ceremonial objects intentionally defaced or toppled.

Many sacred natural sites are in jeopardy. They are subject to a wide range of pressures and threats, external and internal, such as illegal extraction of timber and wildlife, impacts from extractive industries' operations, construction of dams and roads, looting of associated burials and archaeological sites, encroachment by outsiders, disrespectful tourism, degradation of neighbouring environments, and reduction of available lands and resources for traditional peoples.

Environmental changes may also affect sacred natural sites. Careful assessment is needed regarding issues such as fire, which in some cases can cause damage (e.g. Mount Athos, Greece, Kakouros, 2008), while in others can be beneficial if well managed (e.g. Uluru, Calma 2005). It is recommended that protected areas carry out situation analyses of sacred natural sites to record their condition and the threats they face (see Box 9 for summary of Annex 1). This should include the development of plans to reduce natural and human-induced changes. In cases where valuable artefacts occur at sites, improved arrangements for their care may be needed if local conditions of respect and trust change, or new outside influences come to bear.

Many concerns are now being expressed regarding the impacts of climate change as a new threat to biodiversity and the natural environment. Sacred natural sites represent both fragments of habitat that are vulnerable to climate change, as well as elements of potential strategies to mitigate climate change effects. Research into the impacts of climate change on sacred natural sites is an important new area of investigation. One area of mitigation might be to develop lists of vulnerable species and then work with custodian communities to take some of the most threatened species into *ex-situ* conservation associated with the sacred natural site. Achieving this kind of work may require a focused effort and considerable resources, but it may also be more cost-effective than other options.

Box 9. Planning the management of sacred sites in Guatemala

The Guatemala Program of The Nature Conservancy (TNC) has developed a Conservation Area Planning Methodology to improve management of natural and cultural resources in protected areas, including sacred natural sites. TNC has conducted an assessment of the Atitlán Watershed Multiple Use Area, in the highlands of western Guatemala. We concluded, in part:

"The main effects of deterioration identified for sacred sites in Atitlán were landscape deterioration, restriction of access, and general loss of traditional knowledge and culture. The main causes of deterioration identified for sacred sites were deforestation, forest fires, inadequate garbage disposal, lack of recognition of indigenous rights and inadequate management of tourism.

The most important step in the methodology is the identification of the most feasible strategies for the abatement of critical threats to biodiversity and cultural resources and sites.

Using this methodology, the following strategies have been developed during the planning processes conducted in the Atitlán area of Guatemala:

- Research, systematize and disseminate traditional knowledge;
- Enhance cultural sensitivity of staff from government, non-governmental organizations (NGOs) and private companies in indigenous areas;
- Enhance social appreciation of Mayan spiritual guides, and develop a code of rules for visitors to sacred sites;
- The sacredness of a protected area should be made explicit in the decree, management plan interpretation and communications materials produced about it. People should know that the area is considered sacred by local communities and should respect it;
- Strengthen the participation of spiritual guides in the planning and management of protected areas that include sacred sites. We suggest the following ways:
  - Besides including spiritual guides in general meetings, organize special, more in-depth and intimate meetings for them;
  - Even better, ask the local spiritual custodians to include the protection of sacred sites as a special theme in their own meeting and analysis mechanisms;
  - Include them in the formal bodies for the management of the protected area."

For a detailed description of this planning and assessment process, please see Annex 1.

Source: Secaira and Molina, The Nature Conservancy, Guatemala, 2003.

5.8 Desecrations and re-sanctifying

Safeguard against the unintended or deliberate desecration of sacred natural sites and promote the recovery, regeneration and re-sanctifying of damaged sites where appropriate.

The destruction and desecration of sacred natural sites is ongoing and many sites are under threat, though the rate of damage is currently unknown. Sites are damaged for a number of reasons, including competing uses, power struggles, resource pressures, and declining cultural relevance (see Box 1, a summary of the sources of loss of Indian sacred groves). Efforts should be made to better understand the different pressures on sacred natural sites and provide support to prevent damage. If custodians agree, appropriate national lists of sites under threat could be developed to bring attention to their plight and also to stimulate action toward protection and recovery.

There are cases where damage has occurred and sacred natural sites have been re-sanctified. In Mongolia, ceremonies have been held by Buddhist monks to return sites to faith-based management. For many of Mongolia's sacred sites, religious texts exist that describe the nature of the local deity and historic management of the area. Many of these texts have been relocated and are being used for management once again (Case study 13).

Case study 13. The restoration of Mongolia's sacred sites

Mongolia is a traditionally Buddhist country, but during the communist period, from 1924–1989, Buddhism was suppressed in violent purges in which tens of thousands of Buddhist monks were killed, virtually all of Mongolia's 900 monasteries were destroyed, and many sacred Buddhist texts disappeared. By 1989, at the end of the communist era, there was just one working monastery in Ulaanbaatar, the capital city. There was almost nothing left of Mongolia's monastery libraries.

Old solutions to new problems

After the fall of communism in Mongolia, the monasteries were revived, and many of the texts were rediscovered – having been buried or otherwise hidden by monks and lay people. These texts brought an inspiring solution to one of Mongolia's great problems: how to reverse the decline of the rural environment caused by over-grazing, hunting, water shortage, pollution, logging and waste disposal.

Before communism, Mongolians revered their country as one of the most sacred places in the world. Its chosen protector was Avalokitesvara – the deity of compassion – and throughout the land were hundreds of holy mountains and valleys, each with a sacred text explaining why it was sacred, and how to revere it. In 2002, the monks of the newly revived Gandan Monastery, helped by the Alliance for Religion and Conservation, WWF Mongolia and the World Bank, published a collection of ancient *sutras*, or sacred texts, entitled *Sacred Sites in Mongolia*.

Reinstating Mongolia's sacred sites

The texts describe the sacred geography and rituals of 80 sacred sites whose use had been suppressed under communism. So far, seven of the most important Mongolian sacred sites have been re-sanctified and dedicated by their local monastic communities. In Mongolian tradition, sacred landscapes are marked by standing stones, signifying that trees, animals and land in the area should be protected. Even within 12 months, there was a noticeable increase in the number of wild animals, including marmots and birds of prey. It is hoped that in the long term, with nature allowed to take its own course, the vegetation and tree cover will also return.

As part of the process, traditional hunting and logging bans were restored, including a centuries-old ban on hunting the Snow Leopard and the Saiga Antelope, both of which are endangered. The ban is an expression of the ancient Buddhist teaching of compassion toward all life, which in practice encourages Buddhists to engage in sustainable natural resource management.

Among the sacred sites where hunting is now banned is Bogd Khan Mountain, Mongolia's oldest protected area. It dates back to the 12th century, and is the site where hunting bans were first introduced by special decree of the Khan 800 years ago. In June 2003, in response to concerns about the protection of the forested area facing Ulaanbaatar, the Buddhist community unveiled a new carving of the protector deity on the side of the mountain facing the city, in a move intended to strengthen the conservation of the forest. Another region that comes under the full protection of this ban is the Khan Kentii Strictly Protected Area. It encompasses 1.2 million hectares of land and is home to one tenth of Mongolia's forests and many rare and threatened species of plants and animals.

The project has succeeded because the standing stones and what they signify are backed up by the monks in each area, who carry influence with local government and nomadic communities.

Source: The Alliance for Religion and Conservation website (www.arcworld.org), Introduction to *Sacred Sites in Mongolia*, written by Dr Hatgin Sukhbaatar.

5.9 Development pressures

Apply integrated environmental and social impact assessment procedures for developments affecting sacred natural sites and, in the case of the land of indigenous and local communities, support the application of the Convention on Biological Diversity's Akwé: Kon Guidelines for minimizing the impacts of development actions.

Most countries have developed systems of environmental and impact assessment, and include environmental impact statements as part of their development planning legislation. In the case of sacred natural sites it will be important to apply these assessment techniques, ensuring the terms of reference cover a broader range of considerations such as cultural aspects. In some jurisdictions impact assessments are carried out after agreement in principle has already been given, and the scope of the assessment applies only to mitigation of the proposed development. In these situations, mechanisms need to be developed to adequately inform decision makers of the values of the sacred sites at an earlier stage.

The Akwé: Kon Guidelines, drafted by the Convention on Biological Diversity, present a comprehensive set of voluntary procedures for the conduct of cultural and social impact assessments regarding developments proposed to take place on or which are likely to impact on, sacred sites and on lands and waters traditionally occupied or used by indigenous and local communities. Adhering to these guidelines will help achieve accurate assessment of the biological and cultural values of sacred natural sites and these can then be taken into account in the decision-making process. Akwé: Kon (pronounced *akway goo*), is a Mohawk term meaning "everything in creation".[59]

5.10 Financing

Where appropriate, pay due attention to the suitable financing of sacred natural site management and protection, and develop mechanisms for generating and sharing revenue that take into account considerations of transparency, ethics, equity and sustainability. Recognise that in many parts of the world poverty is a cause of the degradation of sacred natural sites.

While issues of financing sacred natural sites can be sensitive, economic values are often integrated into their management in the following ways:

- Sacred natural sites are usually associated with deities, ancestral spirits or saints and other religious figures that are entreated in time of need, including during periods of economic hardship.
- The spiritual leadership of the specific sacred natural site is often charged with the overall well-being of the community as a whole, including their economic welfare.
- Offerings are usually expected in kind or cash on visiting sacred natural sites, similar to temple and church collection boxes, although usually less formalized.
- Custodians are sometimes formally paid or receive donations from their communities to carry out cultural and religious functions.

In addition, sacred natural sites under changing social and economic conditions may require more active management and protection. Increased visitor usage, for example, may require staffing by guides, while economic and changing community dynamics may call for the deployment of rangers. The elders of the Kenya Kayas, for example, felt that lack of livelihood options and poverty were two of the main threats to the long-term survival of the Kayas (see Case study 5).

Adequate consultation with appropriate custodians requires that they travel to meetings and devote valuable time to research, community meetings and dialogue with protected area staff. Plan to reimburse custodians for expenses and compensate them for their time – these gestures are appropriate and will be appreciated.

Understanding the relationships between increasing poverty and declining natural resources, and mitigating both has been a major challenge over the last 30 years. This remains a complex area, particularly with regard to protected areas, which can lock up resources and at the same time provide new livelihood opportunities. Protected areas can preserve resources and livelihoods from further deterioration. Views on poverty vary dramatically from culture to culture and the loss of traditional culture itself is hard to value. Poverty and development debates over the last 50 years have often promoted the view that Western-style development is optimal. This has often implied that many indigenous and local communities are blocks to development. Currently, there is some reassessment of these views given increasing concerns over sustainability, climate change and environmental deterioration.

[59] The Mohawk are an indigenous people of North America.

6 Respect the rights of sacred natural site custodians within an appropriate framework of national policy.

6.1 Institutional analysis

Understand traditional management institutions and enable and strengthen the continued management of sacred natural sites by these institutions. Make appropriate arrangements for the adoption and management of sacred natural sites that have no current custodians, for example by heritage agencies.

Understanding current management institutions: Sacred natural sites have a wide range of traditional institutional arrangements governing them. These are not always understood and recognised by protected area managers and other government agencies. Sometimes the institutional guardians no longer live near the site or key clan members may now live at a very great distance. A better understanding of the institutional arrangements of sacred natural sites and the linking with newer management structures will support improved care and management. Social research and the understanding of common property regimes, as well as of their institutions, have become more developed in recent years. There is now a better understanding of the importance of these institutions and the conservation community has been working with this knowledge to develop co-management options that recognise existing institutions and reflect common property arrangements.

Usually institutions need to make some adjustments to work with protected area agencies. For example, at the Uluru-Kata Tjuta National Park in Australia, a management board has been established which operates under both the *Tjukurpa*, or local law of Anangu people, and the national conservation laws of Australia, so the Western culture of day-to-day park management is guided by Tjukurpa.[60] A similar adjustment has been made at Mount Athos in Greece, where a management board has been created and six ecologist monks have been assigned appropriate research duties (Annex 7).

Management boards are increasingly being used in many protected area contexts to more formally bring stakeholder inputs into protected area management. These can be legally recognised articulations of the ecosystem approach, which calls for stakeholder engagement (guideline 2.7). It is important that protected area agencies develop the flexibility to work with the wide range of institutional arrangements governing sacred natural sites.

Rules are one of the basic elements that define an institution. Custodian institutions have typically developed detailed rules regarding all aspects of the management of sacred natural sites and a common feature of many of these sites is their restricted access and use. When access is permitted, rules typically strongly constrain behaviour. Wherever possible, it will be advantageous to respect and support the custodians' own regulations regarding the management of sacred natural sites, incorporating these into protected area rules as appropriate.

Caring for sites with no current custodians: Sacred natural sites that were historically considered sacred, but for which there are currently no traditional custodians, present a series of difficulties for management. The fact that a site does not have obvious custodians does not mean that local people are not concerned about the site, and many have been looked after informally for long periods of time. It is important to verify that a sacred natural site does not, in fact, have active guardians, as they may exist but not be easily visible.

In some places, government heritage organizations have taken on the care and management of such sites, where they have deemed this to be important. There are also positive examples in which a different spiritual community has taken over the care of a sacred site. For example, Holy Island off the Island of Arran in Scotland is a long neglected Christian sacred natural site that has recently been adopted by a Buddhist community. Some very old sacred sites may currently be managed by government departments responsible for ancient monuments, which may or may not have policies regarding spiritual values.

6.2 Legal protection

Advocate for legal, policy and management changes that reduce human and natural threats to sacred natural sites, especially those not protected within national protected areas and other land planning frameworks.

Sacred natural sites face many threats and many are being lost. They face pressures from all kinds of sources. While those included within legally recognised protected areas should have basic protection, those outside can be threatened significantly. Efforts should be undertaken, with the support of local custodians, to create appropriate legal mechanisms to protect sacred natural sites.

[60] Calma, 2005.

Case study 14. Rila Monastery Natural Park, Bulgaria

The Natural Park of the Monastery of Rila is one of the largest and more significant European protected areas initiated by a Christian Church, in this case the Bulgarian Orthodox Church. It is also a good example of the effective integration of spiritual, cultural and natural values for conservation.

The Rila Monastery was founded by Saint Ivan Rilsky, a hermit, at the beginning of the 10th century. Perhaps due to this origin the "holy unity" between nature and the monastery has always been preserved. The monastery owned and managed extensive properties until 1947, when government nationalization took all of them. Restitution of these properties to the Orthodox Church began in 1998 and continued until 2002. In 1983, UNESCO inscribed Rila Monastery on the list of World Heritage Sites.

For the Bulgarian people, Rila is the holiest place, nestling in the bosom of the most majestic mountain in the country. It is the second largest monastery in the Balkan region, with a millennial history of always being open to people. Several "sanctums" are located around the monastery: holy springs, the holy cave of the founder, five hermitages, and so on. A small monastic community currently serves the monastery.

For centuries the Rila Monastery was a stronghold of the Bulgarian language and culture, a support for the people's consciousness. The first book of Bulgarian history was written in Rila. The monastery has one of the oldest and most important libraries of the Balkan region.

The Natural Park has healthy ecosystems and spectacular mountain landscapes, ranging from 1,000–2,700m, including 28 lakes, most of glacial origin, and outstanding beech, oak, and spruce forests, the oldest in the country. They include 36 native tree species, and many endemic plants, over 20 of which are found only in the area. Fauna diversity is also high, including at the top of the trophic pyramid, healthy populations of wolf and brown bear.

The Natural Park of the Monastery of Rila was established by a Decree of the Ministry of the Environment and Water Resources in the year 2000. The protected area is *ca.* 25,000ha, of which *ca.* 19,000ha belong to the church, and 3,600ha are in state hands. The state property is recognised as a Nature Preserve, and church property is a Natural Park, with a high level of protection. It is surrounded by a National Park.

A management plan for the Natural Park was prepared in 2003 by an interdisciplinary team, with involvement of the management team, and a consultation process. Zoning includes a specific category for religious values. No hunting or harvesting is allowed, except a small quota for the monastic community. The main goals of the management plan include: the conservation of religious and cultural heritage, conservation of natural components, management of natural resources and tourism, interpretation and education, as well as coordination of the activities of the Orthodox Church and state institutions. An important component is to guarantee and preserve the unity between nature and the Rila Monastery, and to support the rebirth of the area as the spiritual and cultural centre of Bulgaria.

Source: National Forestry Board. In the bosom of the Rila Monastery Natural Park. www.bg-parks.net.

6.3 Rights-based approach

Root the management of sacred natural sites in a rights-based approach respecting basic human rights, rights to freedom of religion and worship, and to self-development, self-government and self-determination as appropriate.

In most situations, custodians of different spiritual traditions have found accommodation regarding the use of sacred natural sites that they share (Dudley *et al.*, 2005) (see Case study 11, Sri Pada-Adams Peak). There are, however, situations where practitioners of particular faiths have not been free to worship as they would like, and access to and use of sacred natural sites have been contested. The desecration of sacred natural sites has, in some cases, been used as a tool of domination and to hasten the imposition of external value systems. Freedom of religion and belief is defined as follows in the UN Declaration of Human Rights, adopted by the United Nations General Assembly in 1948:

> "Everyone has the right to freedom of thought, conscience and religion; this right includes freedom to change his religion or belief, and freedom, either alone or in community with others and in public or private, to manifest his religion or belief in teaching, practice, worship, and observance."

In 2007, the United Nations endorsed the Declaration on the Rights of Indigenous Peoples (Annex 3). Box 2 section 3.9 reproduces the key articles and statements related to sacred natural sites.

IUCN also takes a rights-based approach to its work and is concerned with issues of equity as well as of sustainability. The Union's mission is:

> "to influence, encourage and assist societies throughout the world to conserve the integrity and diversity of nature and to ensure that any use of natural resources is equitable and ecologically sustainable".[61]

Protected area staff should receive adequate training and must be expected to respect faiths other than their own.

6.4 Confirm custodians' rights

Support the recognition, within the overall national protected area framework, of the rights of custodians to their autonomous control and management of their sacred sites, and guard against the imposition of conflicting dominant values.

In efforts to improve the conservation of sacred natural sites it is important that the autonomy of their custodians is not compromised. Especially in the case of developing new protected area networks that include sacred natural sites, all efforts should be made to ensure that the management rights of custodians are recognised. For many traditional societies, the future of their communities is bound up in sacred natural sites.

Increasingly, the benefits of protected area status are being recognised by local and indigenous communities. Indigenous groups are proposing that their traditionally conserved areas be recognised by government conservation designations. In Australia, for example, there has been the recent development of "Indigenous Protected Areas" or IPAs (Case study 15). Not only do these Indigenous Protected Areas have significant biological and cultural benefits but they also have significant social benefits (see Box 4).

There remain cases where dominant value systems are being imposed on the custodians of sacred natural sites, sometimes even leading to the destruction or damaging of these sites. In the spirit of the Declaration of Human Rights discussed in guideline 6.3, national governments should make efforts to protect sacred natural sites and their custodians from the imposition of outside value systems.

Case study 15. Dhimurru Indigenous Protected Area (IPA), northeast Arnhem Land, Australia

Dhimurru Indigenous Protected Area (IPA) is located in northeast Arnhem Land, Australia. Located on Aboriginal land, Dhimurru IPA surrounds the township of Nhulunbuy, named after the sacred hill Nhulun at the base of which a mining town is built. Nhulun is a good example of fostering cross-cultural learning and signifies the importance of sacred sites to the land rights movement.

Yolngu people have a cultural responsibility to manage the land in accordance with spiritual obligations to their ancestors. This way of managing the land has sustained Aboriginal peoples' presence on the land since time immemorial. Cultivating a deeper understanding of Yolngu values encompasses the cultural connection to the land which is a sacred one and is secured through the relationships between *rom* (law/protocol), *manikay* (song/ceremony) and *Miny'tji* (art).

Natural and cultural heritage values within Dhimurru IPA are managed from within the cultural worldview of the Yolngu people and therefore are never treated separately. Although the management approach toward the land is always from a Yolngu perspective, Dhimurru fosters different ways of understanding the unique natural heritage in the IPA. To this end, a unique "Both Ways" management approach has been established with the Parks and Wildlife Service of the Northern Territory.

Dhimurru provides an example of a protected area for which indigenous people have sole management responsibility and have chosen to exercise that responsibility by negotiating productive partnerships with government and non-government organizations to produce an alternative form of joint management.

The Yolngu have a rich culture with a deep understanding of nature and they have only been exposed to European values since 1935. It may not come as a surprise to find that most of the impacts on the natural and cultural values of the IPA are known to result from pressures from the outside world. The establishment of an aluminium mine on Yolngu land prompted the concern of Yolngu people for the well-being of their land and their visitors, and subsequently led to the establishment of the Dhimurru Land Management Corporation.

Dhimurru IPA is characterized by unique coastal flora and fauna. Marine turtles and dugong are of particular cultural interest to Yolngu people. Marine turtles are in decline in most parts of the world. Dhimurru has therefore been learning from other monitoring studies to establish good management practices that are known to aid recovery of declining populations. Such

[61] http://www.iucn.org/en/about/

Case study 15. Dhimurru Indigenous Protected Area (IPA), northeast Arnhem Land, Australia (cont.)

management practices include protecting habitat (through Dhimurru's permit system and patrols), advocating the use of TEDs (Turtle Exclusion Devices) in fisheries, and ensuring that the traditional harvest of eggs and turtles is sustainable (through a sea country management plan).

Specific legislation exists to assist Aboriginal people in the Northern Territory to protect and manage their sacred sites. The Aboriginal Areas Protection Authority (AAPA) is a government department responsible for registering and documenting sacred sites reported to them by traditional owners under the 1989 Northern Territory Aboriginal Sacred Sites Act. The AAPA is also responsible for the legal protection of sacred sites and assisting with prosecution in cases in which these sites are impinged upon. Unfortunately, such impingements still occur. Despite being declared an Aboriginal Reserve in 1931 and registration under the Aboriginal Land Rights Act of 1976 (hereafter referred to as ALRA), a general lack of understanding and sometimes respect for Aboriginal culture still persists in the broader Australian society today.

A history of land rights at Nhulun (Nhulun Lookout). Nhulun is a natural hill that is closed in on three sides by Nhulunbuy Township with over 4,000 residents. It was the centre of attention in the first legal claim of Aboriginal traditional ownership of land and customary law that eventually resulted in the establishment of the Aboriginal Land Rights Act (1976). Hence it can be said that the struggle for land rights began as this sacred site was damaged. A protest ceremony known as the *Galtha Bunggul* was held on the damaged site. Subsequently the Yolngu people created a statement of their "title deeds", glued it onto a bark painting (commonly known as "the bark petition") and presented it to the House of Representatives in 1963. The bark petition made special reference to the sacredness of the land.

Traditionally, Nhulun is a sacred site relating to a "Sugarbag" dreaming, the maternal birthplace of the wild honey, or sugarbag, which is produced by native bees and considered a delicacy amongst Yolngu. Nhulun is a registered sacred site under the 1989 Northern Territory Sacred Sites Act, but Nhulun is also a recreational area where residents go for a run, take the dog for a walk or to enjoy the sunset. A sealed road provides access for motorized vehicles to a steel watch tower (Nhulun Lookout) overlooking the surrounding lands. Having had an ultimately successful experience, Dhimurru is aware of the importance of fostering effective ways for achieving reconciliation and cultural understanding.

From the Dhimurru homelands comes the nation's most famous Aboriginal band called Yothu Yindi. The band has been instrumental in bringing issues such as recognition for culture and land rights to the forefront of mainstream Australian society. The band produced songs in Yolngu and English with titles such as "Treaty", "Tribal Voice" and "Mainstream". The band has an international reputation, having toured extensively around the world, establishing relationships with other indigenous peoples' organizations and forming the Yothu Yindi Foundation.

The lookout has been provided with signage on cultural interpretation explaining the cultural and spiritual values of the sacred site as well as its importance to the land rights movement. The signage is a good example of Dhimurru's strategy of promoting reconciliation and cultural understanding through the interpretation of Yolngu beliefs and values to visitors.

Nhulun is of high cultural significance to Yolngu people who still carry out ceremonies at Nhulun. The last ceremony took place on 1 May 2007 and was a re-enactment of the Galtha Bunggul held in 1969.

Despite the legal recognition of Yolngu people's land rights, Yolngu people today still struggle for other people to recognise these rights. To halt illegal access to land and potential damage to sacred sites, Dhimurru manages a permit system that is at the forefront of this fight. Nhulun is an interesting example of the continuous need to educate the public as destruction of the interpretive signage regularly occurs.

Recognising cultural and spiritual connections to land in a legal sense would greatly contribute to resolving concerns that custodians have over the well-being of their sacred sites.

Conclusions

Preserving culture as a whole. Sacred sites are part of a rich cultural tradition intertwining with the land and sea. They can be used to enforce traditional responsibilities in taking care of the land and sea. Cultural concepts such as law, song, art and ceremony should be treated as being part of a living culture.

Empower custodians and people responsible to take appropriate care of their sacred sites. Ensure that local traditional owners are assisted and their concerns addressed through effective legal arrangements. There exists a clear need for Yolngu to keep asserting their rights and educating outsiders. Cultural protocols and intellectual property rights should consider sacred sites and the cultural

Case study 15. Dhimurru Indigenous Protected Area (IPA), northeast Arnhem Land, Australia (cont.)

and spiritual values related to them. This may be important for visiting scientists, commercial operators and companies, as well as other parties from the outside.

Access to sacred sites may be regulated based on zoning and permit schemes. Both may operate concurrently and in synergy with one another. Zoning schemes in particular may be a good way to conserve cultural and natural values.

Appropriate media and communications. Consider media and means of communication that are culturally acceptable and functional. The use of still images and film can be very effective. Such media can also be linked to existing information systems. Communicate aspirations and rights to other organizations and institutions. Promote reconciliation and cultural understanding through the interpretation of Yolngu beliefs and values to visitors. This may be of particular use to establish good working relationships with other parties using the land and sea.

Exchange experiences about the management of cultural and spiritual values as well as sacred sites. Yolngu people have extensive experience in managing sacred sites that may be valuable for other peoples to learn from and *vice versa*.

Authors: Dhimurru Land Management Aboriginal Corporation.

6.5 Tenure

Where sacred natural sites have been incorporated within government or private protected areas in ways that have affected the tenure rights of their custodians, explore options for the devolution of such rights and for their long-term tenure security.

The last 20 years has seen the evolution of shared management responsibilities between conservation agencies and local communities (see Case study 12 Uluru). These have taken the form of collaborative and community management types of governance. These efforts have also been supported by moves to democratize land tenure. For example, reforms in land tenure are taking place in over 30 African states, and many have made efforts toward moving land administration to the local level.[62] These reforms have been instituted to increase efficiency, as well as to empower local citizens and promote democracy. In Tanzania, which has taken the lead in community-based management, a category of "group private" tenure has been established, through which the confirmation of customary village use has occurred. The Tanzanian Forest and Beekeeping Division has promoted community-based forest management across the country. By 2006, over 382 village land forest reserves had been declared or gazetted, covering more than two million hectares of forest and woodland (10.2% of the country's public lands) and involving 1,102 villages.[63] A considerable number of village forest reserves are sacred natural sites or include sacred natural sites within them. These efforts have, therefore, supported the long-term nationwide conservation of sacred natural sites under community management.

A number of centralized economies have also accomplished the return of sacred natural sites from state to community custodial management, including several countries associated with the former Soviet Union (Case study 13, Mongolia; Case study 14, Rila). In Mongolia, for example, the suppression of Buddhism from 1927 to 1989 led to the destruction of hundreds of monasteries and the abandonment of traditional management regimes. Subsequently, numerous sacred natural sites were over-exploited. Efforts have been underway for the last 18 years to re-establish some of the country's estimated 800 sacred natural sites. Progress has been made and many are now back in the hands of local custodians. Some have been endorsed at a national level and have gained the status of "state sacred natural sites".

Communities in many places are increasingly taking steps to activate the recognition and protection of their sacred natural sites.

62 Wily, 2003.
63 FBD, 2006.

Case study 16. Vilcanota Spiritual Park, the first in the world

For the Quechua people of Peru, the glorious 20,945ft (6,372m) Mount Ausangate in the Andes is their main *Apu* or deity. After many years of struggle to protect the sacred mountain, the people finally were able to declare the mountain and its attendant range (Cordillera Vilcanota) as Vilcanota Spiritual Park in December 2004 – the first spiritual park in the world.

Located in southern Peru, near Cusco, the Vilcanota range has 469 glaciers and is the source for a vast watershed system that feeds both the valleys of the high Andes and the Amazonian rainforest below to the east. These waters have sustained an immense diversity of organisms and many great cultures. Mount Ausangate has been sacred to many of Peru's cultures throughout history, including the Incas.

Today, the indigenous Q'eros community of the Quechua people revere the mountains of the Cordillera Vilcanota, believing they are divinities to be protected. As one Quechua shaman, Andres Apaza, says, "The mountains and Mother Earth provide us with life, with crops, with cattle, pasture and shelter. They provide us with the blessing of life". Around 50,000 people make an annual pilgrimage to Ausangate's life-giving glacier each June during the festival of *Q'olloy Rit'I*, "The Star of the Snow". Although no one knows when the ceremony began, with the introduction of Catholicism, people began to carry crosses up the mountain, placing them in the snow to honour both Jesus and the *Apu*.

Led by shamans, the Q'eros have practised rituals for centuries that are equivalent to what we today call adaptive resource management. These practices have successfully protected the ecosystem to this day. However, despite these cultural traditions, the region is threatened by resource exploitation, a warming climate that is melting the glaciers, and the influx of hundreds of thousands of tourists each year. In addition, increased cattle grazing, logging and mining in the fragile Andean highlands have caused a shift in land management from small, local farmers to large outside interests. Local communities are facing pressures to change their livelihoods and their lifestyles to accommodate external forces.

In the face of these challenges, the Quechua, with government support, have established the world's first Spiritual Park, which will directly benefit the communities and their culture, as a formal recognition of the mountainous landscape as a Community Conserved Area with Sacred Natural Sites. The Vilcanota Spiritual Park is significant for its archaeological sites and the mountain massifs that sustain multiple ecosystems over a great portion of South America. It protects a hotspot of biodiversity, including a thousand varieties of native potato.

The Vilcanota Spiritual Park emphasises indigenous and traditional management of conservation and agrobiodiversity activities with a goal of poverty alleviation based on local concepts of well-being (Figures 3 and 4). The park will protect the area's fragile biodiversity. "Because of restricted access to some areas with voluntary protection measures exercised by the local population, Sacred Natural Sites conserve local ecosystems and their unique biodiversity in an effective and efficient way, so they can serve as repositories of critical biological resources for the rehabilitation of depleted Andean landscapes", says Alejandro Argumedo, Director of Asociación Kechua Aymara para Comunidades Sustentables (ANDES), a community-based Quechua-Aymara organization. The traditional management of Vilcanota Spiritual Park is designed to be a model for future special conservation areas in the Peruvian Andes.

Source: Adapted from Sacred Land Film Project, A.Tindall, with thanks to A. Argumedo and K. Mazjoub.

6 Conclusion

It is increasingly recognised that the human species is having a dramatic, negative effect on the planet. We face the erosion of genetic resources, the shrinking of wildlife habitats, the loss of biological diversity at an unprecedented rate, depletion of fisheries and forests, shortages of water, soil and agricultural land, and very worrying disruptions to global climate. At the same time, the world's rich cultural diversity — languages, belief systems, traditional values — are in serious jeopardy. Sacred natural sites represent the meeting ground of cultural and biological diversity, and working collaboratively to understand and protect these sacred places will strengthen the movement to save the planet's priceless biological and cultural mosaic.

The rapid growth of lands and waters designated as legally protected areas is a hopeful indicator of widespread efforts to preserve life on the planet. This modern movement has incorporated the traditional lands of indigenous and other peoples, including their sacred natural sites, into quickly expanding protected area systems. Let us not forget that, in many instances, land and water have been cared for over millennia, and that sacred natural sites and reverence for nature often lie at the core of this ancient, human caretaking impulse.

Sacred natural sites are of concern to most of humanity — from indigenous communities to mainstream faiths, from national governments to secular groups. Sacred natural sites and their associated cultural groups represent deep and diverse relationships with nature, most of which are respectful and life affirming, and they contain seeds that humanity needs to cultivate in order to restore a healthy relationship with the planet. Traditional wisdom articulates a set of community values that calls for human beings to take responsibility for our actions and to protect the places most dear to us.

Protected area managers engaging in respectful partnerships with custodians of sacred natural sites can play a critical role in the survival and effective care of these special places, ensuring the sustainable conservation of both nature and culture for the whole of humanity.

Annex 1

Planning of sacred natural sites in protected areas: an adaption of a methodology for biodiversity and lessons learned in the highlands of western Guatemala

by Estuardo Secaira and Maria Elena Molina, The Nature Conservancy, Guatemala

Summary

A large number of protected areas around the globe were designed for the protection of cultural heritage and biodiversity. The Nature Conservancy (TNC) is committed to the conservation of both, particularly in countries with rich, ongoing indigenous history, such as Guatemala.

The Guatemala Program of TNC has been involved in the large-scale planning of the Maya Biosphere Reserve, a vast expanse of 1.5 million ha of tropical forests and wetlands that contains hundreds of archaeological sites, such as Tikal, Mirador, Piedras Negras and Yaxhà. TNC was subsequently invited by the Ministry of Culture to help with the management plan of Tikal, one of the largest Maya cities and a World Heritage Site.

For this planning effort we adapted TNC's Conservation Area Planning methodology, in order to include tangible cultural targets, which were categorized as cultural regions, areas, zones, sites, structures and removable objects. After this successful adaption, we developed management plans for two municipal parks in the highlands of western Guatemala. In both cases, sacred natural sites were selected as conservation targets by the indigenous inhabitants. A series of workshops with anthropologists were conducted in order to receive feedback for the inclusion of intangible cultural targets, which were categorized as cultural and spiritual values, traditional knowledge and use, social institutions, spirituality (which includes cosmology, ceremonial practices and sacred sites), collective memory, and cultural traditions and habits.

The planning methodology also includes a viability analysis of natural conservation targets, an integrity analysis of tangible cultural targets and a significance analysis of intangible cultural targets. In these analyses, key conservation factors for the long-term survival of the targets are identified, as well as indicators and their ranking. Major threats and opportunities are later identified and prioritized in order to guide the identification and prioritization of strategies.

The development of the management plan consists of a series of workshops and field trips with key community stakeholders. This experience has demonstrated that the inclusion of sacred natural sites is feasible in the context of protected area management, and that the local inhabitants are very interested in the protection of such sites as an inherent component of their natural and cultural heritage.

Introduction

There are an impressive number of protected areas around the globe that were designed for the protection of cultural heritage and biodiversity. The development of methods and tools for the strategic planning of conservation efforts has been a significant contribution of The Nature Conservancy (TNC). However, those methods and tools were originally designed for the conservation of biodiversity.

The Guatemala Program of TNC has been involved in the large-scale planning of the Maya Biosphere Reserve, a series of protected areas and multiple-use zones which aim to protect 1.5 million ha of tropical forests and wetlands teeming with wildlife and hundreds of archaeological sites, some quite significant, such as the Pre-Classic cities of El Mirador and Nakbé, the Classic cities of Tikal, Uaxactún, Piedras Negras, Rio Azul, Yaxhá, Nakum and Naranjo, and the Post-Classic city of Topoxté. When TNC facilitated the process for development of the 2001–2006 Maya Biosphere Reserve Master Plan, two cultural targets were included (archaeological sites and living cultures), along with a complete analysis of biodiversity targets.

As a consequence of the timid inclusion of cultural targets, we were invited by the Ministry of Culture to help with the development of the management plan of Tikal National Park, which protects one of the largest and most impressive Maya cities. The park was the first of only 23 Mixed World Heritage Sites, recognised for both its cultural and natural heritage. For this significant planning effort, we had the challenge of adapting the current TNC Conservation Area Planning methodology for biodiversity to include tangible cultural targets.

In order to conceptualize the adaptations needed to include tangible cultural targets, mainly archaeological sites, our methodology consisted of an exhaustive bibliographic search of planning methodologies and a series of meetings with archaeologists, anthropologists and conservation specialists. The final test was the planning process of the Tikal National Park Master Plan itself, where concepts were fine-tuned and effectively used.

In the meantime, we were also involved in the design of master plans for municipal parks in the highland parks of western Guatemala, where sacred sites and living indigenous cultures are a fundamental part of the landscape.

In order to adapt the methodology for the inclusion of intangible cultural targets, a series of meetings with anthropologists was conducted. In 2005, the development of the 2006–2010 Master Plan for the Atitlán Watershed Multiple Use Reserve, a protected area in western Guatemala, was our most comprehensive attempt to include both tangible and intangible cultural targets, along with biodiversity and economic activities. In the following pages, we explain the methodology and how each of its steps has been adapted for tangible and intangible cultural targets, with an emphasis on sacred natural sites.

Conservation area planning methodology

TNC has developed a Conservation Area Planning (CAP) Methodology as a process to strategically define and prioritize effective actions that abate the most critical threats to biodiversity targets. The process, known as the 5Ss, consists of the following steps (TNC, 2000):

- **Systems**: here we define the biodiversity targets of our conservation efforts, which could be ecological systems, ecological communities, and species or aggregation of species.
- **Viability analysis**: this is an attempt to measure the conservation status of our targets, through the definitions of the key ecological attributes on which the targets depend for their long-term conservation. The ecological attributes are classified in the categories of size, condition and landscape context.
- **Stresses**: this refers to the ecological and cultural stresses from which selected targets are suffering as a result mainly of anthropogenic forces. Stress could be destruction, degradation or fragmentation of habitat.
- **Sources of stress**: this refers to the causes of the ecological stresses that affect the biodiversity targets and cultural stresses affecting sacred natural sites.
- **Strategies**: these consist of the definitions of sets of actions that will be most effective in the abatement of the most critical threats, which are defined as the sources that cause the most serious stress to our targets.
- **Success**: this refers to the selection of a set of indicators that will be most effective in measuring the "success" (or not) of our conservation efforts, referring to the advance of our conservation strategies and the status of our conservation targets.

This CAP methodology emphasises the use of the most complete and up-to-date scientific information in order to produce a solid science-based plan. The process has the flexibility of being conducted either by a small number of technicians in a series of meetings, or it can involve large numbers of people and stakeholders, through workshops and community consultations. The 5Ss scheme, expanded and adapted for cultural targets follows:

- **Cultural targets:** this refers to the cultural assets that justify the establishment of a protected area, either individually or in an association based on characteristics, attributes and/or management needs. Cultural targets have been classified as tangible and intangible.

Tangible cultural targets

The tangible heritage includes physical assets, such as archaeological and historical regions, areas, zones, sites, buildings, and objects from any time in the cultural development of a certain area. A series of categories has been proposed (Molina *et al.*, 2003):

- **Cultural region**: a large-scale geographical region that possesses common cultural characteristics associated by their connectivity, continuity and cultural-historical coherence. For example: the Mesoamerican region, from western Mexico to western Costa Rica, where several cultural groups share the same basic staple crops (corn, beans and squashes), a numerical system based on 20 and a series of very similar solar and lunar calendars.
- **Cultural area**: an association of zones that share more specific cultural features and are linked by historical, ethnological or stylistic aspects. For example: the Maya areas, within the Mesoamerican Region, where all 30 Mayan languages are related to a common origin, and

have grammatical elements and root words in common.

- **Cultural zone**: a geographical unit that had a common government form that resulted in historical and political cohesion in the past. For example: the Piedras Negras and Yaxchilán archaeological zones, where several ancient Mayan sites shared a common ruling dynasty that influenced architectural styles, ceramics etc.
- **Unit/cultural site**: the location of a significant event, an occupation or a prehistoric or historic activity, or a building or structure or association of these, which possess historical, cultural or archaeological value. This may coincide with human settlements (of different sizes), cities, towns, villages, archaeological sites, prehistoric sites, cemeteries or sanctuaries, among others. For example: Tikal, Teotihuacan and Pompeii.
- **Group, sector or grouping of buildings and constructions**: buildings and constructions associated by the historic period in which they were built or the function they played, such as acropolises, squares, archaeological complexes, neighbourhoods or urban centres, religious quarters, living quarters, and production sites, among others. For example: North Acropolis in Tikal or San Francisco de Lima religious complex in Perú.
- **Building/construction**: individual physical works that are related to domestic, civilian, military/defensive, productive, transport and recreational activities, to name a few, and in which human activities took place (houses, temples, palaces, hotels and others). Other structures built to carry out these different functions include bridges, pavements, aqueducts, walls and tunnels.
- **Mobile cultural goods**: relatively small-scale components that may or may not be easily moved. As shown in previous cases, they possess significant historic, artistic, ethnological, palaeontological, archaeological and technological value and are intimately linked to their natural and social surroundings. These objects may include sculptures, stuccoes, stelae, lintels, paintings, utensils, mural paintings and petroglyphs.
- **Historical documents**: all printed or recorded materials, where research and interventions of a given cultural region, area, zone or sites are compiled. These are very important for the knowledge or management of a specific cultural target.

Intangible cultural targets

"Intangible heritage" refers to the practices, representations and expressions, along with knowledge and skills, associated with communities and groups that are recognised as part of their cultural heritage (UNESCO, 2003). Intangible cultural targets are specific practices and representations that have been selected as components that a community or group wants to be protected. In order to better define them, a series of categories, specially associated with nature or a protected area, have been proposed:

- **Intangible values**: the different values that human beings associate with an area or landscape. These values can be spiritual, recreational, related to identity or existence, artistic, aesthetic, educational, scientific (research and monitoring), associated with peace or therapeutic (Harmon and Putney, 2000).
- **Local knowledge and practices related to nature**: the ethno-ecological knowledge that a given cultural group or community has developed over millennia in order to manage natural processes and assure their survival.
- **Social institutions**: the forms of social organization that are the basis for managing and conserving the natural and cultural heritage. These institutions have been successful in conserving many landscapes around the globe, like the customary law system that rules the communal forests of Totonicapán in Guatemala and the sacred groves in Tibet and India.
- **Spirituality**: the practices of several forms of spirituality, from indigenous cosmologies to several of the organized religions, give a fundamental value to conservation and are conducted in natural sites, such as the summits of volcanoes and mountains, caves, and the shores of rivers and lagoons. Many of these sacred natural sites are located in protected areas. Stewardship of these sites is often of crucial importance to nearby local communities, and consequently, protected area managers need to recognise the importance of shared responsibility for management.
- **Collective memory**: the historical facts or mythical accounts that form part of the cultural heritage of communities and peoples around the globe. Those facts and accounts refer to specific sites, some of which are today located within protected areas.
- **Traditions**: this broad category comprises all cultural practices not specified above, which have an intrinsic relationship with the landscape or area that is to be protected. These can include many cultural attributes such as language, music, dances, theatre, gastronomy, festivals and traditional games.

Examples of natural, tangible and intangible cultural targets are shown here for the Atitlán Watershed Multiple Use Area, in the highlands of western Guatemala:

Biodiversity targets:

- broadleaved forest
- mixed forest (pine and oak)
- high-altitude conifer forests
- xerophytic association
- Lake Atitlán watershed
- Madrevieja and Nahualate watershed
- horned guan

Cultural targets:

- Mayan archaeological sites
- colonial, republican and vernacular architecture
- Mayan, colonial and republican movable objects
- sacred sites
- oral history, traditional knowledge and practices
- community values and organization
- scenic and recreational values

Analysis of conservation status

The original methodology includes a viability analysis of natural targets, which seeks to define the likelihood of a biodiversity target persisting over the long run (about 100 years). In this analysis, key ecological attributes for the long-term survival of the target are identified, as well as indicators and ranking ranges for qualification.

Through the adaption of the methodology, parallel concepts were defined for cultural heritage.

For tangible cultural targets, the categories of analysis have been defined as follows:

Table 1

Viability analysis	Integrity analysis	Significance analysis	Significance analysis
Natural targets	*Tangible cultural targets*	*Intangible cultural targets*	*Sacred sites*
Size	Conceptual meaning	Correspondence	Correspondence
Condition	Physical condition	Transmissibility	Physical condition
Landscape context	Social and natural context	Context	Context

- **Conceptual content**: the extent to which an element reflects the sociocultural values of the historical period from which it originates, its authenticity, age, information and meaning it transmits.
- **Physical condition**: comparison between an element's original and its current state, based on:
  - how intact it is compared with its original state;
  - how fragmented it is (extension, volume, number of architectural elements);
  - how altered it is on a spatial level by changes, justified and non-justified attachments, stratification, etc.;
  - how degraded its materials and shapes are.
- **Context**: based on the natural and social surroundings, this includes natural and/or social factors that contribute or impinge upon the conservation or degradation of selected cultural targets.

For intangible cultural targets, the categories are defined as:

- **Correspondence**: the degree to which the intangible cultural target is functional for extant communities and groups, and the degree to which it corresponds with the ideology that gave it origin.

Perhaps its significance has been altered, but is still valid for current communities.

- **Transmissibility**: the existence of effective mechanisms for the transmission of knowledge and practices of the intangible cultural target.
- **Context**: the contextual factors that contribute, or not, to the conservation of the intangible cultural target. Among the most significant factors are the legal, institutional and social frameworks on which the intangible cultural targets depend.

For sacred sites, practice has shown that they are a combination of a tangible and an intangible cultural target, and consequently four categories are necessary for their analysis: correspondence, physical condition, transmissibility and context.

Table 2

Rank	Concept
Very good	The indicator is functioning within its optimal range of variation, and does not require intervention for its maintenance
Good	The indicator is functioning within its acceptable range of variation, although it may require some intervention for maintenance
Fair	Range of acceptable variation, and requires intervention for maintenance
Poor	The indicator is far from its acceptable range of variation, and allowing it to remain in this condition for an extended period of time will make restoration or prevention of extirpation of the target practically impossible

Table 3

Category	Indicator	Poor	Fair	Good	Very good
Physical condition	Percentage of sacred sites that conserve their natural context and are clean	Less than 30%	30–59%	60–89%	90% or more
Correspondence	Percentage of sacred sites that show evidence of recent ceremonial use	Less than 30%	30–59%	60–89%	90% or more
Transmissibility	Percentage of participants in Mayan ceremonies at the sacred sites who are less than 40 year old	Less than 10%	11–30%	31–50%	More than 50%
Context	Percentage of sacred sites that receive formal protection	Less than 25%	25–50%	51–75%	More than 75%

An indicator should be identified for each factor, and then a qualification needs to be identified according to the ranks shown in Table 2.

In Table 3, the significance analysis of the sacred sites of the Atitlán Watershed Multiple Use Reserve is shown.

Threat analysis

This refers to the identification of the most critical threats that are causing destruction or degradation of the conservation targets. In order to perform a better, more objective analysis, the threats are analysed in two components:

- **Stresses**: the ecological effects on the target. A stress to a biodiversity target can be habitat destruction or degradation. For cultural targets, specialists prefer to call them deterioration effects. The stresses or effects of deterioration are ranked by the criteria of severity and scope.
- **Sources of stress**: the human or natural causes of the stress. Sources of stress can be the advance of the agricultural frontier, overfishing, incompatible forestry or cattle-ranching practices, infrastructure development, mining etc. For cultural targets, the same concept has been framed as causes of deterioration. The sources of stress or causes of deterioration are ranked by the criteria of contribution and irreversibility.

The main effects of deterioration identified for sacred sites in Atitlán were landscape deterioration, restriction of access, and general loss of traditional knowledge and culture. The main causes of deterioration identified for sacred sites were deforestation, forest fires, inadequate garbage disposal, lack of recognition of indigenous rights and inadequate management of tourism.

Strategies

The most important step in the methodology is the identification of the most feasible strategies for the abatement of critical threats to biodiversity and cultural targets. The strategies are then ranked by the following criteria: benefits (in terms of attaining objectives, abating threats or improving viability), feasibility and costs.

Using this methodology, the following strategies have been developed during the planning processes conducted in the Atitlán area of Guatemala:

- Research, systematize and disseminate traditional knowledge.
- Enhance cultural sensitivity of staff from government, non-governmental organizations (NGOs) and private companies in indigenous areas.

- Enhance social appreciation of Mayan spiritual guides, and develop a code of rules for visitors to sacred sites.

Conclusions and recommendations

The process for the development of management plans has consisted of a series of workshops with representatives of the different stakeholders in the community, including elders and religious leaders. Some aspects of the analysis are conducted through field trips, where the planning team goes collectively to the field and conducts an *in situ* analysis of what needs to be done in order to conserve the area.

This experience has demonstrated that the inclusion of sacred natural sites is feasible in the context of protected areas, and that the local inhabitants are very interested in the protection of such sites as an inherent component of their natural and cultural heritage. Out of this experience we can make the following conclusions and suggestions:

Sacred natural sites have started to be included systematically in the planning and management of protected areas in the highlands of Guatemala.

We suggest that this approach should be included in a revised version of the national guidelines for the elaboration of protected area management plans.

The sacredness of a protected area should be made explicit in the decree, management plan interpretation and communications materials produced about it. People should know that the area is considered sacred by local communities and should respect it.

However, we need to strengthen the participation of spiritual guides in the planning and management of protected areas that include sacred sites. We suggest the following way:

Besides including spiritual guides in general meetings, organize special, more in-depth and intimate meetings for them.

Even better, ask the local spiritual custodians to include the protection of sacred sites as a special theme in their own meeting and analysis mechanisms.

Include them in the formal bodies for the management of the protected area.

References and further reading

Harmon, D. and Putney, A. 2000. "The Non-Material Values of Protected Areas". *Parks Magazine*, Vol. 10, No.2, IUCN.

Molina, M.E., Secaira, E., Lehnhoff, A., Martin, A, Chan, R.M., Vaverde, M.J., Edwards, M. 2003. *Conservation Area Planning for Tangible Cultural Resources.* Guatemala, The Nature Conservancy.

TNC. 2000. *Esquema de las 5 S's para la Conservación de Sitios. Guatemala,* The Nature Conservancy. (In Spanish.) Online: http://www.nature.org/aboutus/howwework/cbd/science/art14309.html

UNESCO. 2003. *Convention on Intangible Cultural Heritage.* Paris, UNESCO.

Annex 2

Extract from the guidelines for image capture at Uluru-Kata Tjuta National Park, Australia

Guidelines for commercial image capture, use and commercial sound recording

Uluru-Kata Tjuta National Park is a World Heritage Area renowned for its exceptional natural environment and the living culture of its traditional Aboriginal owners, known as 'Anangu'. Anangu and Parks Australia welcome filmmakers, photographers and painters to the Park. We ask you to work with us to help respect Anangu culture and protect this unique living cultural landscape and the fragile desert ecosystems for future generations.

Anangu own the Park and lease it to the Director of National Parks, who manages it through Parks Australia. The lease requires Parks Australia to take steps to safeguard Anangu traditions.

These guidelines have been developed to help you carry out your work. They have been drawn up in collaboration with traditional owners and representatives from the tourism, film and photographic industries. The guidelines are a compromise between protection of Anangu culture and the requirements of visitors. They have been agreed to in the spirit of joint management and are approved by the Park's Board of Management.

The guidelines will help you have a better understanding of the Park's World Heritage values.

The Park's Media Office is available to answer any questions you may have. You will find useful information at the Parks web site (www.deh.gov.au/parks/uluru/index.html) and at Attachment A, which includes an explanation of many of the terms we use …

Why Are Permits Needed?

Protecting Cultural Values

2.1 Uluru-Kata Tjuta National Park's cultural significance comes from Anangu traditions dating back tens of thousands of years. The foundation of Anangu culture is Tjukurpa, which is the source of stories, ceremonies, landscapes, plants and animals, art and rules for living (www.deh.gov.au/parks/uluru/tjukurpa/index.html). The international significance of this living culture was recognised in 1994 when the Park became one of the first World Heritage 'cultural landscapes'.

Permits help us to protect Anangu culture. Under Tjukurpa, certain sites, ritual objects, designs and ceremonies are restricted to people who may 'properly' view them. Some stories and sites are restricted to initiated men, others to women. Some stories may be spoken, but not written or filmed. Anangu believe that their culture is degraded if images of sacred sites are captured or displayed, or if the Park is used to advertise products and services that do not promote the natural and cultural values of the Park, and people's awareness, understanding and enjoyment of them.

2.2 The capture and use of pictures and names of Anangu is also sensitive. Like most people, Anangu do not like being filmed or photographed unless they have given their permission. Also, under Tjukurpa, Anangu honour their dead by not speaking their names or looking at their pictures for some considerable time. For this reason we ask that you seek permission each time you wish to publish pictures of Anangu …

Working In The Park

Uluru has great spiritual significance to Anangu. There are many sensitive sites associated with Uluru itself. We ask that you respect these places and their significance to Anangu by following these guidelines. Map 1 shows the parts of Uluru and specific sites that should not be filmed, photographed or painted, and locations from which images should not be captured. Most sites are clearly signposted, although some are not as they cannot be specifically identified or publicly discussed. If you are unsure whether you have inadvertently captured an image of a sensitive site, you can ask us to look at any images you are planning to use.

3.1 Please do not capture images of sensitive sites as indicated by S on Map 1 – Uluru

3.2 Please be careful when capturing images from the sunset viewing area as you may be capturing images of sensitive sites.

3.3 Please advise the Park Media Office if you want to capture images of the northeast face of Uluru as it is not generally allowed. This is because there are many sacred sites across this face, some of which we cannot show on Map 1 for cultural reasons. Where approval is given, sacred sites must be obscured (for example by shadow, a bush or sand dune) or not in clear view. We will advise you on locations and times of day to help you get the shots you need whilst respecting Anangu culture. All images of the northeast face must be submitted for approval prior to use (see 4.3 – 4.9).

3.4 Anangu prefer that people don't climb Uluru and do not want pictures promoting the climb. Please do not capture images which focus on the climb such as people climbing, the base of the climb, the chain on the climb or views from the climb, including images from the top of Uluru.

3.5 Please do not capture images of rock art unless given specific approval by Parks Australia. Approval will require consultation with Anangu and accompaniment by Anangu representatives when carrying out your work.

Kata Tjuta is a sacred area. Under Anangu law details of the stories cannot be revealed and access to some areas is restricted. We ask that you respect these sacred places by following these guidelines. Map 2 – Kata Tjuta shows the parts of Kata Tjuta that may not be filmed, photographed or painted.

3.6 For cultural reasons, no filming, photography or artwork is allowed along the Valley of the Winds walking track.

3.7 You may carry out panoramic filming, photography or artwork, in other publicly accessible areas in and around Kata Tjuta but images must include three complete domes, never focusing on single domes. This is to make sure you do not focus on sacred detail (see photographic insert example in Map 2).

3.8 You may film, paint, draw or photograph within Walpa Gorge provided both sides of the Gorge are in frame to avoid revealing sacred places (see photographic insert example in Map 2.) ...

Some definitions:

Cultural values: Anangu cultural values are diverse and complex and not easily understood by Western systems of classification. For Anangu, the Park is part of a living landscape. Features like Uluru, vegetation, soils and animals are all vested with cultural meaning in ways that transcend Western separation of "nature" and "culture", people and landscapes and notions of past, present and future. Cultural meaning and values are expressed through language, song, story, art and an array of cultural beliefs and practices that are governed by "Tjukurpa". The Park was World Heritage listed in 1994 as an outstanding example of traditional human land use and being directly associated with living traditions and beliefs of outstanding universal significance.

Sacred sites: a site that is sacred to Anangu or is otherwise of significance to Anangu according to Anangu tradition. Anangu law may restrict detailed knowledge of sacred sites to particular people who are responsible for particular sites. Knowledge of sacred sites may not be public knowledge.

Sensitive sites: the sites shown on Map 1 and other sacred sites that can't be shown or are not publicly known for cultural reasons.

For further information

http://www.deh.gov.au/parks/uluru/vis-info/permits.html

Uluru-Kata Tjuta National Park
Media Office
PO Box 119, Yulara, NT 0872

Phone: (08) 89561113 Fax: (08) 89562360
Email: uluru.media@deh.gov.au

Source

http://www.environment.gov.au/parks/uluru/vis-info/permits-image.html

Map 1 Uluru

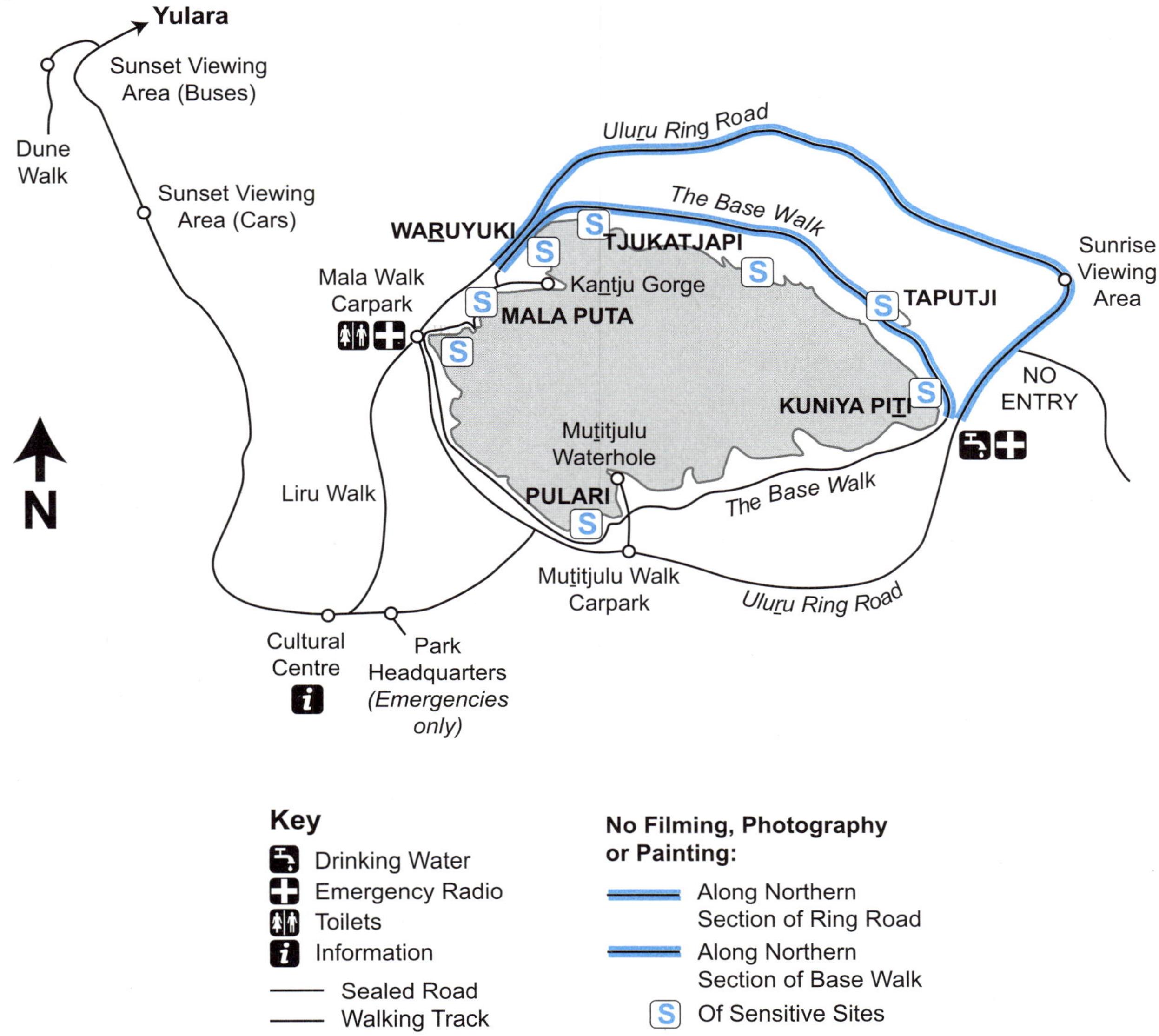
Yulara
Sunset Viewing Area (Buses)
Dune Walk
Sunset Viewing Area (Cars)
Uluru Ring Road
The Base Walk
WARUYUKI
TJUKATJAPI
Mala Walk Carpark
Kantju Gorge
MALA PUTA
TAPUTJI
Sunrise Viewing Area
NO ENTRY
KUNIYA PITI
Mutitjulu Waterhole
Liru Walk
PULARI
The Base Walk
N
Mutitjulu Walk Carpark
Uluru Ring Road
Cultural Centre
Park Headquarters (Emergencies only)
Key
Drinking Water
Emergency Radio
Toilets
Information
Sealed Road
Walking Track
No Filming, Photography or Painting:
Along Northern Section of Ring Road
Along Northern Section of Base Walk
Of Sensitive Sites

Annex 3

UN Declaration on the Rights of Indigenous Peoples, 2007

Resolution adopted by the General Assembly

[*without reference to a Main Committee (A/61/L.67 and Add.1)*]

61/295. United Nations Declaration on the Rights of Indigenous Peoples

The General Assembly,

Taking note of the recommendation of the Human Rights Council contained in its resolution 1/2 of 29 June 2006,[1] by which the Council adopted the text of the United Nations Declaration on the Rights of Indigenous Peoples,

Recalling its resolution 61/178 of 20 December 2006, by which it decided to defer consideration of and action on the Declaration to allow time for further consultations thereon, and also decided to conclude its consideration before the end of the sixty-first session of the General Assembly,

Adopts the United Nations Declaration on the Rights of Indigenous Peoples as contained in the annex to the present resolution.

107th plenary meeting
13 September 2007

Annex
United Nations Declaration on the Rights of Indigenous Peoples

The General Assembly,

Guided by the purposes and principles of the Charter of the United Nations, and good faith in the fulfilment of the obligations assumed by States in accordance with the Charter,

Affirming that indigenous peoples are equal to all other peoples, while recognizing the right of all peoples to be different, to consider themselves different, and to be respected as such,

Affirming also that all peoples contribute to the diversity and richness of civilizations and cultures, which constitute the common heritage of humankind,

Affirming further that all doctrines, policies and practices based on or advocating superiority of peoples or individuals on the basis of national origin or racial, religious, ethnic or cultural differences are racist, scientifically false, legally invalid, morally condemnable and socially unjust,

Reaffirming that indigenous peoples, in the exercise of their rights, should be free from discrimination of any kind,

Concerned that indigenous peoples have suffered from historic injustices as a result of, inter alia, their colonization and dispossession of their lands, territories and resources, thus preventing them from exercising, in particular, their right to development in accordance with their own needs and interests,

Recognizing the urgent need to respect and promote the inherent rights of indigenous peoples which derive from their political, economic and social structures and from their cultures, spiritual traditions, histories and philosophies, especially their rights to their lands, territories and resources,

Recognizing also the urgent need to respect and promote the rights of indigenous peoples affirmed in treaties, agreements and other constructive arrangements with States,

Welcoming the fact that indigenous peoples are organizing themselves for political, economic, social and cultural enhancement and in order to bring to an end all forms of discrimination and oppression wherever they occur,

Convinced that control by indigenous peoples over developments affecting them and their lands, territories and resources will enable them to maintain and strengthen their institutions, cultures and traditions, and to promote their development in accordance with their aspirations and needs,

Recognizing that respect for indigenous knowledge, cultures and traditional practices contributes to sustainable and equitable development and proper management of the environment,

---

[1] See *Official Records of the General Assembly, Sixty-first Session, Supplement No. 53* (A/61/53), part one, chap. II, sect. A

Emphasizing the contribution of the demilitarization of the lands and territories of indigenous peoples to peace, economic and social progress and development, understanding and friendly relations among nations and peoples of the world,

Recognizing in particular the right of indigenous families and communities to retain shared responsibility for the upbringing, training, education and well-being of their children, consistent with the rights of the child,

Considering that the rights affirmed in treaties, agreements and other constructive arrangements between States and indigenous peoples are, in some situations, matters of international concern, interest, responsibility and character,

Considering also that treaties, agreements and other constructive arrangements, and the relationship they represent, are the basis for a strengthened partnership between indigenous peoples and States,

Acknowledging that the Charter of the United Nations, the International Covenant on Economic, Social and Cultural Rights[2] and the International Covenant on Civil and Political Rights,[2] as well as the Vienna Declaration and Programme of Action,[3] affirm the fundamental importance of the right to self-determination of all peoples, by virtue of which they freely determine their political status and freely pursue their economic, social and cultural development,

Bearing in mind that nothing in this Declaration may be used to deny any peoples their right to self-determination, exercised in conformity with international law,

Convinced that the recognition of the rights of indigenous peoples in this Declaration will enhance harmonious and cooperative relations between the State and indigenous peoples, based on principles of justice, democracy, respect for human rights, non-discrimination and good faith,

Encouraging States to comply with and effectively implement all their obligations as they apply to indigenous peoples under international instruments, in particular those related to human rights, in consultation and cooperation with the peoples concerned,

Emphasizing that the United Nations has an important and continuing role to play in promoting and protecting the rights of indigenous peoples,

Believing that this Declaration is a further important step forward for the recognition, promotion and protection of the rights and freedoms of indigenous peoples and in the development of relevant activities of the United Nations system in this field,

Recognizing and reaffirming that indigenous individuals are entitled without discrimination to all human rights recognized in international law, and that indigenous peoples possess collective rights which are indispensable for their existence, well-being and integral development as peoples,

Recognizing that the situation of indigenous peoples varies from region to region and from country to country and that the significance of national and regional particularities and various historical and cultural backgrounds should be taken into consideration,

Solemnly proclaims the following United Nations Declaration on the Rights of Indigenous Peoples as a standard of achievement to be pursued in a spirit of partnership and mutual respect:

Article 1

Indigenous peoples have the right to the full enjoyment, as a collective or as individuals, of all human rights and fundamental freedoms as recognized in the Charter of the United Nations, the Universal Declaration of Human Rights[4] and international human rights law.

Article 2

Indigenous peoples and individuals are free and equal to all other peoples and individuals and have the right to be free from any kind of discrimination, in the exercise of their rights, in particular that based on their indigenous origin or identity.

Article 3

Indigenous peoples have the right to self-determination. By virtue of that right they freely determine their political status and freely pursue their economic, social and cultural development.

Article 4

Indigenous peoples, in exercising their right to self-determination, have the right to autonomy or self-government in matters relating to their internal and local affairs, as well as ways and means for financing their autonomous functions.

Article 5

Indigenous peoples have the right to maintain and strengthen their distinct political, legal, economic, social and cultural institutions, while retaining their right to participate fully, if they so choose, in the political, economic, social and cultural life of the State.

---

2 See resolution 2200 A (XXI), annex.
3 A/CONF.157/24 (Part I), chap. III.
4 Resolution 217 A (III).

Article 6

Every indigenous individual has the right to a nationality.

Article 7

1. Indigenous individuals have the rights to life, physical and mental integrity, liberty and security of person.
2. Indigenous peoples have the collective right to live in freedom, peace and security as distinct peoples and shall not be subjected to any act of genocide or any other act of violence, including forcibly removing children of the group to another group.

Article 8

1. Indigenous peoples and individuals have the right not to be subjected to forced assimilation or destruction of their culture.
2. States shall provide effective mechanisms for prevention of, and redress for:

   (a) Any action which has the aim or effect of depriving them of their integrity as distinct peoples, or of their cultural values or ethnic identities;

   (b) Any action which has the aim or effect of dispossessing them of their lands, territories or resources;

   (c) Any form of forced population transfer which has the aim or effect of violating or undermining any of their rights;

   (d) Any form of forced assimilation or integration;

   (e) Any form of propaganda designed to promote or incite racial or ethnic discrimination directed against them.

Article 9

Indigenous peoples and individuals have the right to belong to an indigenous community or nation, in accordance with the traditions and customs of the community or nation concerned.

Article 10

Indigenous peoples shall not be forcibly removed from their lands or territories. No relocation shall take place without the free, prior and informed consent of the indigenous peoples concerned and after agreement on just and fair compensation and, where possible, with the option of return.

Article 11

1. Indigenous peoples have the right to practise and revitalize their cultural traditions and customs. This includes the right to maintain, protect and develop the past, present and future manifestations of their cultures, such as archaeological and historical sites, artefacts, designs, ceremonies, technologies and visual and performing arts and literature.
2. States shall provide redress through effective mechanisms, which may include restitution, developed in conjunction with indigenous peoples, with respect to their cultural, intellectual, religious and spiritual property taken without their free, prior and informed consent or in violation of their laws, traditions and customs.

Article 12

1. Indigenous peoples have the right to manifest, practise, develop and teach their spiritual and religious traditions, customs and ceremonies; the right to maintain, protect, and have access in privacy to their religious and cultural sites; the right to the use and control of their ceremonial objects; and the right to the repatriation of their human remains.
2. States shall seek to enable the access and/or repatriation of ceremonial objects and human remains in their possession through fair, transparent and effective mechanisms developed in conjunction with indigenous peoples concerned.

Article 13

1. Indigenous peoples have the right to revitalize, use, develop and transmit to future generations their histories, languages, oral traditions, philosophies, writing systems and literatures, and to designate and retain their own names for communities, places and persons.
2. States shall take effective measures to ensure that this right is protected and also to ensure that indigenous peoples can understand and be understood in political, legal and administrative proceedings, where necessary through the provision of interpretation or by other appropriate means.

Article 14

1. Indigenous peoples have the right to establish and control their educational systems and institutions providing education in their own languages, in a manner appropriate to their cultural methods of teaching and learning.
2. Indigenous individuals, particularly children, have the right to all levels and forms of education of the State without discrimination.
3. States shall, in conjunction with indigenous peoples, take effective measures, in order for indigenous individuals, particularly children, including those living outside their communities, to have access, when possible, to an education in their own culture and provided in their own language.

Article 15

1. Indigenous peoples have the right to the dignity and diversity of their cultures, traditions, histories and aspirations which shall be appropriately reflected in education and public information.

2. States shall take effective measures, in consultation and cooperation with the indigenous peoples concerned, to combat prejudice and eliminate discrimination and to promote tolerance, understanding and good relations among indigenous peoples and all other segments of society.

Article 16

1. Indigenous peoples have the right to establish their own media in their own languages and to have access to all forms of non-indigenous media without discrimination.

2. States shall take effective measures to ensure that State-owned media duly reflect indigenous cultural diversity. States, without prejudice to ensuring full freedom of expression, should encourage privately owned media to adequately reflect indigenous cultural diversity.

Article 17

1. Indigenous individuals and peoples have the right to enjoy fully all rights established under applicable international and domestic labour law.

2. States shall in consultation and cooperation with indigenous peoples take specific measures to protect indigenous children from economic exploitation and from performing any work that is likely to be hazardous or to interfere with the child's education, or to be harmful to the child's health or physical, mental, spiritual, moral or social development, taking into account their special vulnerability and the importance of education for their empowerment.

3. Indigenous individuals have the right not to be subjected to any discriminatory conditions of labour and, inter alia, employment or salary.

Article 18

Indigenous peoples have the right to participate in decision-making in matters which would affect their rights, through representatives chosen by themselves in accordance with their own procedures, as well as to maintain and develop their own indigenous decision-making institutions.

Article 19

States shall consult and cooperate in good faith with the indigenous peoples concerned through their own representative institutions in order to obtain their free, prior and informed consent before adopting and implementing legislative or administrative measures that may affect them.

Article 20

1. Indigenous peoples have the right to maintain and develop their political, economic and social systems or institutions, to be secure in the enjoyment of their own means of subsistence and development, and to engage freely in all their traditional and other economic activities.

2. Indigenous peoples deprived of their means of subsistence and development are entitled to just and fair redress.

Article 21

1. Indigenous peoples have the right, without discrimination, to the improvement of their economic and social conditions, including, inter alia, in the areas of education, employment, vocational training and retraining, housing, sanitation, health and social security.

2. States shall take effective measures and, where appropriate, special measures to ensure continuing improvement of their economic and social conditions. Particular attention shall be paid to the rights and special needs of indigenous elders, women, youth, children and persons with disabilities.

Article 22

1. Particular attention shall be paid to the rights and special needs of indigenous elders, women, youth, children and persons with disabilities in the implementation of this Declaration.

2. States shall take measures, in conjunction with indigenous peoples, to ensure that indigenous women and children enjoy the full protection and guarantees against all forms of violence and discrimination.

Article 23

Indigenous peoples have the right to determine and develop priorities and strategies for exercising their right to development. In particular, indigenous peoples have the right to be actively involved in developing and determining health, housing and other economic and social programmes affecting them and, as far as possible, to administer such programmes through their own institutions.

Article 24

1. Indigenous peoples have the right to their traditional medicines and to maintain their health practices, including the conservation of their vital medicinal plants, animals and minerals. Indigenous individuals also have the right to access, without any discrimination, to all social and health services.

2. Indigenous individuals have an equal right to the enjoyment of the highest attainable standard of physical and mental health. States shall take the necessary steps with a view to achieving progressively the full realization of this right.

Article 25

Indigenous peoples have the right to maintain and strengthen their distinctive spiritual relationship with their traditionally owned or otherwise occupied and used lands, territories, waters and coastal seas and other resources and to uphold their responsibilities to future generations in this regard.

Article 26

1. Indigenous peoples have the right to the lands, territories and resources which they have traditionally owned, occupied or otherwise used or acquired.
2. Indigenous peoples have the right to own, use, develop and control the lands, territories and resources that they possess by reason of traditional ownership or other traditional occupation or use, as well as those which they have otherwise acquired.
3. States shall give legal recognition and protection to these lands, territories and resources. Such recognition shall be conducted with due respect to the customs, traditions and land tenure systems of the indigenous peoples concerned.

Article 27

States shall establish and implement, in conjunction with indigenous peoples concerned, a fair, independent, impartial, open and transparent process, giving due recognition to indigenous peoples' laws, traditions, customs and land tenure systems, to recognize and adjudicate the rights of indigenous peoples pertaining to their lands, territories and resources, including those which were traditionally owned or otherwise occupied or used. Indigenous peoples shall have the right to participate in this process.

Article 28

1. Indigenous peoples have the right to redress, by means that can include restitution or, when this is not possible, just, fair and equitable compensation, for the lands, territories and resources which they have traditionally owned or otherwise occupied or used, and which have been confiscated, taken, occupied, used or damaged without their free, prior and informed consent.
2. Unless otherwise freely agreed upon by the peoples concerned, compensation shall take the form of lands, territories and resources equal in quality, size and legal status or of monetary compensation or other appropriate redress.

Article 29

1. Indigenous peoples have the right to the conservation and protection of the environment and the productive capacity of their lands or territories and resources. States shall establish and implement assistance programmes for indigenous peoples for such conservation and protection, without discrimination.
2. States shall take effective measures to ensure that no storage or disposal of hazardous materials shall take place in the lands or territories of indigenous peoples without their free, prior and informed consent.
3. States shall also take effective measures to ensure, as needed, that programmes for monitoring, maintaining and restoring the health of indigenous peoples, as developed and implemented by the peoples affected by such materials, are duly implemented.

Article 30

1. Military activities shall not take place in the lands or territories of indigenous peoples, unless justified by a relevant public interest or otherwise freely agreed with or requested by the indigenous peoples concerned.
2. States shall undertake effective consultations with the indigenous peoples concerned, through appropriate procedures and in particular through their representative institutions, prior to using their lands or territories for military activities.

Article 31

1. Indigenous peoples have the right to maintain, control, rotect and develop their cultural heritage, traditional knowledge and traditional cultural expressions, as well as the manifestations of their sciences, technologies and cultures, including human and genetic resources, seeds, medicines, knowledge of the properties of fauna and flora, oral traditions, literatures, designs, sports and traditional games and visual and performing arts. They also have the right to maintain, control, protect and develop their intellectual property over such cultural heritage, traditional knowledge, and traditional cultural expressions.
2. In conjunction with indigenous peoples, States shall take effective measures to recognize and protect the exercise of these rights.

Article 32

1. Indigenous peoples have the right to determine and develop priorities and strategies for the development or use of their lands or territories and other resources.
2. States shall consult and cooperate in good faith with the indigenous peoples concerned through their own representative institutions in order to obtain their free and informed consent prior to the approval of any project affecting their lands or territories and other resources, particularly in connection with the development, utilization or exploitation of mineral, water or other resources.
3. States shall provide effective mechanisms for just and fair redress for any such activities, and appropriate measures shall be taken to mitigate adverse environmental, economic, social, cultural or spiritual impact.

Article 33

1. Indigenous peoples have the right to determine their own identity or membership in accordance with their customs and traditions. This does not impair the right of indigenous individuals to obtain citizenship of the States in which they live.
2. Indigenous peoples have the right to determine the structures and to select the membership of their institutions in accordance with their own procedures.

Article 34

Indigenous peoples have the right to promote, develop and maintain their institutional structures and their distinctive customs, spirituality, traditions, procedures, practices and, in the cases where they exist, juridical systems or customs, in accordance with international human rights standards.

Article 35

Indigenous peoples have the right to determine the responsibilities of individuals to their communities.

Article 36

1. Indigenous peoples, in particular those divided by international borders, have the right to maintain and develop contacts, relations and cooperation, including activities for spiritual, cultural, political, economic and social purposes, with their own members as well as other peoples across borders.
2. States, in consultation and cooperation with indigenous peoples, shall take effective measures to facilitate the exercise and ensure the implementation of this right.

Article 37

1. Indigenous peoples have the right to the recognition, observance and enforcement of treaties, agreements and other constructive arrangements concluded with States or their successors and to have States honour and respect such treaties, agreements and other constructive arrangements.
2. Nothing in this Declaration may be interpreted as diminishing or eliminating the rights of indigenous peoples contained in treaties, agreements and other constructive arrangements.

Article 38

States in consultation and cooperation with indigenous peoples, shall take the appropriate measures, including legislative measures, to achieve the ends of this Declaration.

Article 39

Indigenous peoples have the right to have access to financial and technical assistance from States and through international cooperation, for the enjoyment of the rights contained in this Declaration.

Article 40

Indigenous peoples have the right to access to and prompt decision through just and fair procedures for the resolution of conflicts and disputes with States or other parties, as well as to effective remedies for all infringements of their individual and collective rights. Such a decision shall give due consideration to the customs, traditions, rules and legal systems of the indigenous peoples concerned and international human rights.

Article 41

The organs and specialized agencies of the United Nations system and other intergovernmental organizations shall contribute to the full realization of the provisions of this Declaration through the mobilization, inter alia, of financial cooperation and technical assistance. Ways and means of ensuring participation of indigenous peoples on issues affecting them shall be established.

Article 42

The United Nations, its bodies, including the Permanent Forum on Indigenous Issues, and specialized agencies, including at the country level, and States shall promote respect for and full application of the provisions of this Declaration and follow up the effectiveness of this Declaration.

Article 43

The rights recognized herein constitute the minimum standards for the survival, dignity and well-being of the indigenous peoples of the world.

Article 44

All the rights and freedoms recognized herein are equally guaranteed to male and female indigenous individuals.

Article 45

Nothing in this Declaration may be construed as diminishing or extinguishing the rights indigenous peoples have now or may acquire in the future.

Article 46

1. Nothing in this Declaration may be interpreted as implying for any State, people, group or person any right to engage in any activity or to perform any act contrary to the Charter of the United Nations or construed as authorizing or encouraging any action which would dismember or impair, totally or in part, the territorial integrity or political unity of sovereign and independent States.

2. In the exercise of the rights enunciated in the present Declaration, human rights and fundamental freedoms of all shall be respected. The exercise of the rights set forth in this Declaration shall be subject only to such limitations as are determined by law and in accordance with international human rights obligations. Any such limitations shall be non-discriminatory and strictly necessary solely for the purpose of securing due recognition and respect for the rights and freedoms of others and for meeting the just and most compelling requirements of a democratic society.

3. The provisions set forth in this Declaration shall be interpreted in accordance with the principles of justice, democracy, respect for human rights, equality, non-discrimination, good governance and good faith.

Annex 4

IUCN Resolution on Indigenous Peoples and Protected Areas, 1996

IUCN World Conservation Congress (WCC)

Resolution 1.53 – Indigenous Peoples and Protected Areas (October 1996)

RECALLING that some protected areas have been established on indigenous lands and territories without the consent and participation of the affected people;

CONSIDERING the terms of ILO Convention No 169 and those of the Convention on Biological Diversity, regarding the role of indigenous peoples with respect to the management, use and conservation of biodiversity;

CONSIDERING the recommendations and guidelines established in Agenda 21;

CONSIDERING the emphasis placed in *Caring for the Earth* on the role of indigenous peoples in sustainable development and their rights in the management of natural resources;

CONSIDERING the recommendations of the IVth World Congress on National Parks and Protected Areas, calling for the development of policies for protected areas which safeguard the interests of indigenous peoples;

RECOGNIZING that several governments have already adopted policies and measures to fully incorporate the rights and interests of indigenous peoples in the establishment and management of protected areas within their lands and territories;

The World Conservation Congress at its 1st Session in Montreal, Canada, 14–23 October 1996:

1. REQUESTS the Director General, the Secretariat and technical programmes, Commissions, members and Councillors of IUCN, within available resources, to endorse, support, participate in and advocate the development and implementation of a clear policy in relation to protected areas established in indigenous lands and territories, based on the following principles:

   (a) recognition of the rights of indigenous peoples with regard to their lands or territories and resources that fall within protected areas;

   (b) recognition of the necessity of reaching agreements with indigenous peoples prior to the establishment of protected areas in their lands or territories;

   (c) recognition of the rights of the indigenous peoples concerned to participate effectively in the management of the protected areas established on their lands or territories, and to be consulted on the adoption of any decision that affects their rights and interests over those lands or territories;

2. URGES all IUCN members to establish appropriate mechanisms at the national level, for the development and implementation of policies on protected areas and indigenous peoples that are consistent with these principles.

3. REQUESTS the World Commission on Protected Areas to establish closer links with indigenous people's organizations, with a view to incorporating the rights and interests of indigenous peoples in the application of the IUCN Protected Area Management Categories.

4. REQUESTS the Director General, within available resources, to incorporate in IUCN's work on protected areas and natural heritage, specific actions aimed at ensuring the further development and implementation of appropriate policies based on these principles.

Annex 5

Definition of indigenous and local communities

A. ILO definition of indigenous and tribal peoples

C169 Indigenous and Tribal Peoples Convention, 1989

1. This Convention applies to:

(a) tribal peoples in independent countries whose social, cultural and economic conditions distinguish them from other sections of the national community, and whose status is regulated wholly or partially by their own customs or traditions or by special laws or regulations;

(b) peoples in independent countries who are regarded as indigenous on account of their descent from the populations which inhabited the country, or a geographical region to which the country belongs, at the time of conquest or colonisation or the establishment of present state boundaries and who, irrespective of their legal status, retain some or all of their own social, economic, cultural and political institutions.

2. Self-identification as indigenous or tribal shall be regarded as a fundamental criterion for determining the groups to which the provisions of this Convention apply (Article 1).

B. Definition of local communities (from Borrini-Feyerabend *et al.*; 2004)

Local communities

A community is a human group sharing a territory and involved in different but related aspects of livelihoods—such as managing natural resources, producing knowledge and culture, and developing productive technologies and practices. Since this definition can apply to a range of sizes (is a city a community? is the sum of all people inhabiting a watershed a community?), it can be further specified that the members of a "*local* community" are those people that are likely to have *face-to-face* encounters and/or *direct* mutual influences in their daily life. In this sense, a rural village, a clan in transhumance or the inhabitants of an urban neighbourhood can be considered a "local community", but not all the inhabitants of a district, a city quarter or even a rural town. A local community could be permanently settled or mobile.

Most communities have developed their identity and cultural characteristics over time by devising and applying a strategy to cope with a given environment and manage its natural resources. They possess a distinctive form of social organization, and their members share in varying degrees political, economic, social and cultural characteristics (in particular language, behavioural norms, values, aspirations and often also health and disease patterns). They also function, or have functioned in the past, as micro-political bodies with specific capacities and authority.

Important processes in community life comprise social integration (cooperation to address common needs), social conflict (clashing of needs and wants among individual members or families within the community), cultural continuity and cultural change. Mechanisms that generally promote integration in communities include patterns of reciprocity (such as exchanges in marriages or economic trade) and redistribution (sharing economic surpluses among individuals or families). Conditions that may promote conflict in communities include major differences in power and status, e.g. among the young and the elderly, men and women, or among different community units (households) or sub-groups (clans, classes, occupational groups, castes, interest associations, etc.). Such differences are usually reflected in different access to resources (land, capital, water, trees, services, etc.), sometimes leading to exploitation (getting more than others in a common activity), accumulation (avoiding the sharing of surpluses) and the possible splitting or break-down of communities.

In order to survive and develop as a social body, a community continually manages a balance between the opposing forces of conflict and integration, continuity and change. The capacity of a community to deal with these phenomena through time can be used as a criterion

Annex 6

V^th IUCN World Parks Congress, 2003: Extracts from Durban Accord and WPC Recommendation on Cultural and Spiritual Values

We, the 3000 participants of the Vth IUCN World Parks Congress, celebrate, voice concern and call for urgent action on protected areas. We bear witness to those places most inspirational and spiritual, most critical to the survival of species and ecosystems, most crucial in safeguarding food, air and water, most essential in stabilising climate, most unique in cultural and natural heritage and therefore most deserving of humankind's special care. We urge action for the benefit of protected areas so that their benefits may be conserved and equitably shared.

Cause for celebration

We celebrate the miracle of the diversity of nature and of cultures that possess the wealth, the wisdom and the knowledge to enable conservation and sustainable use.

And we celebrate the conservation successes of local communities, indigenous peoples, governments, private individuals and voluntary organisations, and their efforts to make protected areas places of natural, cultural and spiritual convergence.

Cause for concern

We voice concern that many areas of irreplaceable and immediately threatened biological diversity have not yet been protected.

We voice concern that many places conserved over the ages by local communities, mobile and indigenous peoples are not given recognition, protection and support.

We voice concern that wild and natural areas outside of protected areas have shrunk by half in the last 20 years, and that biological diversity, in turn, is on the brink of mass extinction.

Call for commitment and action

We urge commitment to expand and strengthen worldwide systems of protected areas, prioritised on the basis of imminent threats to biodiversity, natural and cultural heritage, and with special attention to marine and freshwater ecosystems.

We urge commitment to involve local communities, indigenous and mobile peoples in the creation, proclamation and management of protected areas.

We urge commitment to protected area management that strives to reduce, and in no way exacerbates, poverty.

We urge commitment to protected area management that shares benefits with indigenous peoples, mobile peoples and local communities.

We urge commitment to recognise, strengthen, protect and support community conservation areas.

Our Pledge

Our strongest commitments will fail if we neglect to maintain avenues for open dialogue. Such dialogue thrives in a climate of humility, credibility and trust. Towards this end we pledge to facilitate understanding and collaboration. We pledge to engage and embrace all constituencies. We pledge to share our vision that a sustainable future for humankind depends on a caring partnership with nature. We pledge to bequeath protected areas, as a precious heritage, to future generations.

WPC Recommendation V.13 on cultural and spiritual values of protected areas

The establishment of protected areas is the result of conscious choices of human societies to conserve nature, biodiversity and areas of special cultural value and significance.

Individuals and communities often use protected areas for spiritual reasons, because they inspire and heal them and/or provide them with a place for peace, education and communion with the natural world.

Many transboundary protected areas have already been promoted and managed as areas for peace and cooperation, thus adding a tangible and valuable dimension of peace-building among peoples, nations and communities.

Protected areas serve as fundamental tools for conservation of nature, and thus are an expression of the highest desires and commitments of humankind for the preservation of life on the planet, and that as such, those areas constitute places of deep reverence and ethical realisation.

Many societies, especially indigenous and traditional peoples, recognise sacred places and engage in traditional practices for the protection of geographical areas, nature, ecosystems, or species, as an expression of societal or cultural choice and of their worldview of the sacredness of nature and its inextricable links with culture. They also recognise sacred places as a unique source of knowledge and understanding of their own culture thus providing what could be considered the equivalent of a university.

Sacred places are revered and cared for by indigenous and traditional peoples and are a fundamental part of their territories, bringing significant benefits to local, national, and global communities.

In some cases, they are seeking to have them recognised as part of existing protected areas systems.

With these points in mind participants in the Session entitled 'Building cultural support for protected areas' held in the Building Broader Support Workshop Stream, recommended that all protected area systems, recognise and incorporate spiritual values of protected areas and culture-based approaches to conservation.

Therefore, PARTICIPANTS in the Stream on Building Broader Support for Protected Areas at the Vth IUCN World Parks Congress in Durban, South Africa (8–17 September 2003):

1. ACKNOWLEDGE indigenous peoples' internationally guaranteed rights to, among others, own and control their sacred places, their archaeological and cultural heritage, ceremonial objects and human remains contained in museums or collections within or adjacent to protected areas. These include the following rights to:

   a. Define and name their sacred places and objects, ancestral remains and archaeological, cultural and intellectual heritage and to have such designations respected as authoritative;

   b. Where relevant, maintain secrecy about and enjoy privacy in relation to their heritage, objects, remains and places as described above;

   c. Receive restitution of sacred places, heritage, objects and remains taken without their free and informed consent;

   d. Freely exercise their ceremonies, religious and spiritual practices in the manner to which they are accustomed;

   e. Gather, collect or harvest flora, fauna and other natural resources used in ceremonies and practices that take place at sacred places or places of archaeological and cultural heritage; and

   f. Maintain their responsibilities to their ancestors and future generations;

2. THEREFORE RECOMMEND that international institutions, governments, protected area authorities, NGOs, churches, user and interest groups fully recognise and respect the abovementioned rights in relation to conservation activities;

3. RECOMMEND that governments should:

   a. PROMOTE and ADOPT laws and policies that foster multicultural values and approaches to protected area systems;

   b. PROMOTE and ADOPT laws and policies that acknowledge the importance of sacred places, particularly those of indigenous and traditional peoples, as valuable for biodiversity conservation and ecosystem management;

   c. ADOPT and ENFORCE laws and policies, with the full and effective participation and consent of peoples and communities concerned, which protect the integrity of sacred places;

   d. ADOPT and ENFORCE laws and policies that guarantee the restitution of sacred places as well as effective control and decision-making processes by local communities and indigenous peoples;

   e. PROMOTE and ADOPT laws and policies, which recognise the effectiveness of innovative governance models such as Community Conserved Areas of indigenous peoples and local communities to ensure control and adequate protection over sacred areas;

   f. PROMOTE and IMPLEMENT effective action to support community protection efforts in areas of cultural and spiritual importance including sacred places; and

g. ADOPT and ENFORCE policies and legal measures, which respect customary use and management of sacred places and ensure access for traditional practitioners in protected areas;

4. FURTHER RECOMMEND that governments, NGOs, local communities and civil society should:

a. ENSURE that protected area systems, protected area designation, objective setting, management planning, zoning and training of managers, especially at the local level, give balanced attention to the full spectrum of material, cultural and spiritual values;

b. ASSIST indigenous and traditional peoples in obtaining legal and technical support related to protection of their sacred places when requested and in a manner that respects their rights and interests; and

c. DEVELOP and IMPLEMENT public education and media campaigns to raise awareness and respect for cultural and spiritual values and, in particular, sacred places;

5. REQUEST protected area managers to:

a. IDENTIFY and RECOGNISE sacred places within their protected areas, with the participation and informed consent of those who revere such places, and to involve them actively in decisions regarding management and protection of their sacred places;

b. PROMOTE intercultural dialogue and conflict resolution with indigenous peoples, local communities and other actors interested in conservation;

c. SUPPORT the efforts of such communities to maintain their cultural and spiritual values and practices related to protected areas; and

d. PROMOTE the use of indigenous languages in these matters;

6. RECOGNISING the importance of cultural and spiritual values in all protected area categories, REQUEST IUCN to review the 1994 Protected Area Category Guidelines with the aim of including these values as additional potential management objectives in categories where they are currently excluded; and

7. REQUEST the IUCN World Commission on Protected Areas and its members to plan and implement actions within the protected areas component of the IUCN Programme for supporting the application of the actions recommended above.

Annex 7

Montserrat and Ouranoupolis Statements of the Delos Initiative

The Montserrat Statement on sacred natural sites in technologically developed countries

TAKING INTO ACCOUNT the papers and case studies from three continents presented during the Montserrat Workshop on Nature and Spirituality, held on 23–26 November 2006 in the Monastery of Montserrat in Catalonia, Spain, in the framework of the IUCN/WCPA Delos Initiative, part of the Task Force on Cultural and Spiritual Values of Protected Areas, as well as the discussions that followed;

BENEFITING from the knowledge and experience of the 40 participants from eight countries, including representatives of Christian and Buddhist monasteries, international, national and local environmental authorities and conservation organizations, academics and individual experts;

APPRECIATING the work carried out by the Delos Initiative during the past two years on sacred natural sites in technologically developed countries;

REALIZING that sacred natural sites are under threat even in developed countries from ignorance and neglect, and specifically from cultural or spiritual breakdown, unsustainable development projects and resource exploitation, urbanization, mass tourism and lack of appropriate land use planning and control;

THE PARTICIPANTS OF THE WORKSHOP:

MAINTAIN that the sacred has been one of the most powerful drivers or realities for conservation, inspiring feelings of awe, veneration and respect. Sacred natural sites, landscapes, species, or particular elements have been one of the most effective forms of nature conservation over the ages, some of them being of local significance, whereas others have significance for wider groups, cultures, traditions, and regions;

RECOGNISE that nature has intrinsic values and meanings, including cultural and spiritual, and is understood by followers of various faiths and spiritual traditions as a divine manifestation of some deeper, sacred reality, however that may be conceived;

FURTHER RECOGNISE that for assurance of long-term sustainability, conservation goals, programmes and messages need to be grounded in deeply held values, beliefs, ideas and practices. The conservation community needs to recognise these aspects and give these deeply held values, beliefs, ideas and practices the place that they deserve in the conservation of protected areas. This constitutes both a challenge and a great opportunity to build further support for the conservation movement, involving partners and stakeholders that up to the present have not been supportive, because they felt excluded by the materialistic outlook that nature conservation has often adopted;

RECOGNISE AND CONFIRM the actual existence of sacred natural sites in all of the IUCN categories of protected areas found in technologically developed countries;

FURTHER CONFIRM that the spiritual aspects of sacred sites in protected areas can contribute significantly to the conservation of natural heritage in various ways, mainly through raising awareness of faithful, inspiring people and involving them in conservation initiatives;

INSIST that the diachronic rights of the custodians of sacred sites must be safeguarded both from insensitive public and private development and from political ignorance and that their participation in determining the future of protected areas must be ensured;

ACKNOWLEDGE that the appropriate management of protected areas can enhance the maintenance of the spiritual values of sacred natural sites, thus creating positive synergy among natural, cultural and spiritual values not only for protected sites, but for nature in general;

FURTHER ACKNOWLEDGE that positive synergies between natural, cultural and spiritual values extend to sacred sites beyond the boundaries of designated Protected Areas and therefore function as a vehicle for supporting and communicating nature conservation;

MAINTAIN that such synergy can be established only through close and equitable collaboration between the traditional custodians of sacred sites and those charged with the management of protected areas, with mutual respect of the prerogatives and responsibilities of the two sides;

SUGGEST that dialogue between these two sides should be encouraged and strengthened, so that objectives and requirements can be fully understood and a common language and – eventually – a common approach, which integrates both views, can be developed;

FURTHER SUGGEST that a future goal would be the development and implementation of integrated management plans for sacred natural sites, which would take into account both spiritual and nature conservation goals; such plans, however, must be prepared with full participation of all those concerned;

POINT OUT that any management measures concerning cultural, spiritual and natural values in sacred natural sites must respect the universal rights of people and be based on a broad participatory approach;

PLEAD FOR CAUTION that particular care should be devoted to sites in protected areas that are sacred for indigenous and traditional peoples and for minorities; in addition, the role and contribution of new immigrants must be taken into account;

SUGGEST that in places where multiple faiths coexist, a diversity of perspectives must be recognised, and the collaboration between these faiths on nature conservation, emphasising shared values, must be promoted;

INDICATE that education and art can play a key role in creating a climate of co-operation and understanding in regard to sacred natural sites;

ENCOURAGE the Delos Initiative to continue its work in facilitating integration processes between the spiritual-faith groups and the nature conservation organizations and consequently develop and disseminate guidance on the integrated management of sacred natural sites;

FINALLY EXPRESS THEIR GRATITUDE to the Monastery of Montserrat, the Board of the Mountain of Montserrat, the Ministry of Environment and Housing of the Government of Catalonia, the Fundació Territori i Paisatge of Catalunya Savings Bank, and the Delos Co-ordination for the excellent organization of the Montserrat Workshop.

The Ouranoupolis Statement[1] on sacred natural sites in technologically developed countries

TAKING INTO ACCOUNT the inspiring message of HAH the Ecumenical Patriarch Bartholomew to the Workshop participants;

NOTING the positive content of the welcoming addresses to the Delos 2 Workshop by the representatives of the Holy Community of Mt. Athos and of the Prefecture of Halkidiki;

CONSIDERING the contributions to the Workshop of the 22 participants from 11 countries in the form of case studies, papers and discussions;

EVALUATING the work carried out in the framework of the Delos Initiative during the past year;

THE PARTICIPANTS OF THE WORKSHOP arrived at the following conclusions:

Lessons from the case studies presented at the Workshop

All parties involved in sacred natural sites within protected areas should be encouraged to resolve in an equitable manner legal and land and resource ownership and use problems, as a prerequisite to achieving collaboration and synergy.

A rights-based approach and the empowerment of local communities should be promoted in all cases.

Through appropriate methods, the mutual awareness, sensitivity and capacity of both custodians of sacred sites and protected area managers – as well as other interested parties – should be cultivated, leading to a common approach for conserving all relevant natural, cultural and spiritual values.

In this context, the establishment of joint approaches of research and learning among the key sides should be promoted.

The custodians of sacred sites should be fully involved in the development of management plans for protected areas, as well as participation in their management structures, and their contribution should be recognised and encouraged.

The right of custodians of sacred sites to limit access to certain areas should be respected and the appropriate measures

[1] Conclusions of a Delos Initiative Workshop, held in Ouranoupolis (Greece) on 24–28 October 2007, in the framework of IUCN/WCPA and its Task Force on Cultural and Spiritual Values of Protected Areas.

should be included in management planning and public use, together with restrictions to visitor access in highly sensitive natural areas.

In addition, the PARTICIPANTS OF THE WORKSHOP agreed on the following specific points:

- The efforts of the Native American nations that are fighting to protect the Holy San Francisco Peaks from ski resort development should be supported.
- The Government of South Korea and the Convention on Wetlands should be encouraged to recognise the Mani-san tidal flats as a Ramsar wetland of international importance, thus contributing to their conservation.
- The efforts to designate the Solovetsky World Heritage Site both for culture (as it is already since 1992) and for nature should be encouraged, thus recognising its integrated character.
- The efforts of the Holy Convent of Chryssopighi in Chania, Crete, to protect its surrounding natural and cultural area from tourist and development pressures should be strengthened with the collaboration of local authorities.

International conventions and organizations could play a positive role in supporting internationally recognised sacred natural sites and should be encouraged to intensify their efforts for this purpose.

Management of monastic lands

Managers of protected areas, as well as relevant policy-makers, must be encouraged to respect the sanctity of Monastic sites and lands under their care and take them into consideration in planning, management and evaluation.

Monastic communities must also be encouraged to manage their lands and facilities in an ecologically sustainable and equitable manner.

For this purpose, a joint and integrated approach to management planning and evaluation is required, which should be developed with participation from both sides.

Monastic sacred sites can provide useful inputs to education and public awareness and their contribution should be encouraged in related protected areas.

In addition, the principles and practice of Monastic asceticism may provide profound lessons towards ecological life styles and the sustainable use of natural resources – as demonstrated by the experience of the Holy Convent of Chryssopighi.

In this context, the positive initiatives of the Holy Community of Mt. Athos in preparing a special environmental study and a strategic plan for the entire autonomous territory, as well as measures for its implementation, through the establishment of a management body under the Holy Monastic Community, with the support of competent scientists and government agencies, were noted with appreciation.

Guidance for sacred natural sites in developed countries

Efforts should be strengthened to sensitize mainstream faiths – both authorities and followers – to nature conservation and broader environmental issues.

The potential contribution of protected natural areas to the conservation of sacred sites and *vice versa* should be enhanced.

Particular respect and care should be addressed to indigenous sacred sites in developed countries, which should be safeguarded from insensitive development pressures.

The Delos Initiative towards WCC[2]

Priority should be given to the publication of the proceedings of the Ouranoupolis Workshop, together with the proceedings of the Montserrat Workshop, for dissemination during the WCC, as a contribution of the Delos Initiative to the ongoing dialogue on natural and cultural / spiritual values and to the implementation of conservation approaches that are sensitive to the cultures and values of all communities.

The Delos Initiative should contribute to the ongoing UNESCO/IUCN process for preparing guidance on sacred natural sites to be completed in 2008. In addition, the Delos Initiative should continue its systematic work for developing further guidance during the triennium 2008–2010 on the specificities of sacred natural sites in developed countries mainly for mainstream religions.

Additional case studies of sacred natural sites should be analysed in the framework of the Initiative, with the goals to obtain more balanced geographical distribution, greater representativity of faiths and the recognition of successful examples, which can disseminate best practices and inspire emulation.

Naturally, the Delos Initiative is expected to contribute substantially to the appropriate events of the 2008 World Conservation Congress, within the framework of the WCPA Task Force on Cultural and Spiritual Values of Protected Areas.

The Initiative should examine potential synergy in its work with related international conventions and organizations, and especially with the World Heritage Convention.

[2] IUCN Fourth World Conservation Congress, Barcelona, Catalonia (Spain), October 2008.

Appreciation

Profound recognition is due of the broad and creative role played by HAH the Ecumenical Patriarch Bartholomew in bringing together faiths in favour of the Creation and the conservation of the natural environment, and sincere gratitude is expressed for his message of advice to the Delos 2 Workshop.

Appreciation is also due of the official participation in the Workshop of the Holy Community of Mt. Athos, of the Greek Ministry of Environment, Physical Planning and Public Works and of the Prefecture of Halkidiki (Northern Greece).

The hospitable and highly informative reception of participants in the Sacred Monasteries of Varlaam in Meteora (Thessaly) and of the Assumption in Ormylia (Halkidiki) is also appreciated.

Special thanks are addressed to Med-INA (the Mediterranean Institute for Nature and Anthropos) for having organized the workshop with efficiency and sensitivity and having covered its costs – with the support of the A. G. Leventis Foundation.

Annex 8

Internet and other resources

Communication, Education and Public Awareness

Convention on Biological Diversity: http://www.cbd.int/cepa/resources.shtml
IUCN Commission on Education and Communication: http://cec.wcln.org
Sacred Land Film Project: http://www.sacredland.org – produced the award-winning documentary film on sacred natural sites, *In the Light of Reverence*, which is distributed by: http://www.bullfrogfilms.com/catalog/ilr.html

Community Conserved Areas

IUCN Best Practice Protected Area Guidelines No. 11 – *Indigenous and Local Communities and Protected Areas* (contains a full chapter on CCAs): http://www.iucn.org/themes/ceesp/Publications/TILCEPA/guidelinesindigenouspeople.pdf
TILCEPA briefing note on CCAs:
http://www.iucn.org/themes/ceesp/wkg_grp/tilcepa/CCA%20Briefing%20Note.pdf
More material on CCAs: www.tilcepa.org

Conflict management in natural resources and generally

IDRC: http://www.idrc.ca/en/ev–28105–201–1-DO_TOPIC. html
Harvard Program on Negotiation: http://www.pon.harvard.edu/research/projects/hnrp.php3

Conservation, ecology and religion

Alliance of Religion and Conservation (ARC): http://www.arcworld.org
Forum on Religion & Ecology: http://religionandecology.org
Delos Initiative: http://www.med-ina.org/delos/

Ecosystem approach

Commission for Ecosystem Management: http://www.iucn.org/themes/cem/ourwork/ecapproach/index.html

Ethnobotany and plant assessment

People and Plants International: http://peopleandplants.org

Free, prior and informed consent

Forest Peoples Programme: http://www.forestpeoples.org/documents/law_hr/fpic_synthesis_jun07_eng.pdf

Indigenous knowledge

Convention on Biological Diversity: http://www.cbd.int/tk/

Intangible cultural heritage

UNESCO Oral and Intangible Heritage of Humanity: http://www.unesco.org/culture/intangible-heritage/

Participatory mapping and participatory GIS

Integrated Approaches to Participatory Development (IPAD): http://www.iapad.org/

Poverty and environment

Poverty Conservation and Learning Group: http://www.povertyandconservation.info/

Millennium Ecosystem Assessment

http://www.millenniumassessment.org

World Database on Protected Areas

http://sea.unep-wcmc.org/wdbpa/

References

Adams, J.S. and McShane, T.O. 1992. *The Myth of Wild Africa: Conservation without illusion.* London, UK: W.W. Norton and Co.

Barnes, 2003. "Wilderness as contested ground". In Harmon and Putney, *op. cit.*

Barrow, E. and Pathak, N. 2005. "Conserving 'unprotected' protected areas – communities can and do conserve landscapes of all sorts". In Brown *et al., op. cit.*

Beltrán, J. 2000. *Indigenous and Traditional Peoples and Protected Areas: Principles, Guidelines and Case Studies.* Best Practice Protected Area Guidelines Series No. 4. Gland, Switzerland and Cambridge, UK: IUCN and WWF International.

Berkes, F. 1999. *Sacred Ecology: Traditional ecological knowledge and resource management.* Philadelphia, USA: Taylor & Francis.

Bhagwat, S. and Rutte, C. 2006. "Sacred Groves: potential for biodiversity management". *Frontiers of Ecology and the Environment* 4(10): 519–524.

Boucher, C. 2002a. *Digging our Roots: the Chamare Museum Frescoes.* Malawi: KuNgoni.

Boucher, C. 2002b. *The Gospel Seed: culture and faith in Malawi as expressed in the Misso Banner.* Malawi: KuNgoni.

Borrini-Feyerabend, G., Kothari, A. and Oviedo, G. 2004. *Indigenous and Local Communities and Protected Areas: Towards Equity and Enhanced Conservation.* Guidance on policy and practice for Co-managed Protected Areas and Community Conserved Areas. Best Practice Protected Area Guidelines Series No. 11. Gland, Switzerland and Cambridge, UK: IUCN.

Borrini-Feyerabend, G., MacDonald, K.I. and Maffi, L. (Eds). 2004. "History, Culture and Conservation". *Policy Matters* 13. IUCN CEESP.

Borrini-Feyerabend, G., Pimbert, M., Favar, M.T., Kothari, A. and Renard, Y. 2004. *Sharing Power: Learning-by-doing in Co-management of Natural Resources throughout the World.* CENESTA, Tehran, Iran: IIED and IUCN/CEESP/CMWG.

Brown, J., Mitchell, N. and Beresford, M. (Eds). 2005. *The Protected Landscape Approach: Linking Nature, Culture and Community.* Gland, Switzerland and Cambridge, UK: IUCN.

Bryceson, D.F and Fonseca, F. 2005. "Risking death for survival: peasant responses to hunger and HIV/AIDS". Paper presented at the International Conference on HIV/AIDS, IFPRI South Africa 14–16 April 2005, Durban SA.

Buckles, D. (Ed.) 1999. *Cultivating Peace: conflict and collaboration in natural resource management.* International Development Research Centre (IDRC)/World Bank Institute.

Chape, S., Harrison, J., Spalding, M. and Lysenko, I. 2005. "Measuring the extent and effectiveness of protected areas as an indicator for meeting global biodiversity targets". *Phil. Trans. R. Soc. B.* 360: 443–455.

Chape, S., Blyth, S., Fish, L., Fox, P. and Spalding, M. (Compilers). 2003. *2003 United Nations List of Protected Areas.* Gland, Switzerland and Cambridge, UK: IUCN and UNEP-WCMC.

Chatterjee, S.,. Gokhale, Y., Malhotra, K.C. and Srivastava, S. 2004. "Sacred Groves in India: an overview". In Dudley *et al., op.cit.*

Colchester, M. and Ferrari, M.F. 2007. *Making FPIC – free, prior and informed consent work: challenges and prospects for Indigenous Peoples.* Moreton-in-Marsh, UK: Forest People Programme.

Colfer, C.J.P. 1995. Who counts most in sustainable forest management? CIFOR Working Paper no.7. http://www.cifor.cgiar.org/publications/pdf_files/WPapers/WP–07.pdf

Cordell, J. 1992. "Who owns the land? Indigenous involvement in Australian Protected Areas". In Kempf, 1992. *op. cit.*

Cunningham, A.B. 2000. *Applied Ethnobotany: people, wild plant use and conservation.* People and Plants Conservation Series. London, UK: Earthscan.

De Lacy, T. and Lawson, B. 1997. "The Uluru/Kakadu model: Joint management of Aboriginal-owned national parks in Australia". In Stevens, 1997, *op. cit.*

Drobyshev, Yu. I., Bazha, S.N., Gunin, P.D., Dugarjav, Ch. and Prischepa, A.V. 2007. "The role of sacred objects in nature conservation of the Baikal Lake basin". UNESCO, 2007, *op.cit.*

Dudley, N., Higgins-Zogib L. and Mansourian, S. 2005. "Beyond Belief, Linking faiths and protected areas to support biodiversity conservation". A research report by WWF, Equilibrium and The Alliance of Religions and Conservation (ARC). Available at: http://assets.panda.org/downloads/beyondbelief.pdf

Eade, D. 2002. *Development and Culture.* London, UK: Oxfam.

Edwards, J. and Palmer, M. 1997. *Holy ground: the guide to faith and ecology.* Northamptonshire, UK: Pilkington Press.

Elias, D. 2003. "Sacred sites in the Tanami Desert, Central Australia". In UNESCO, 2003b, *op. cit.*

FBD. 2006. *Participatory Forest Management in Tanzania: Facts and Figures.* Dar es Salaam, Tanzania: Forest and Beekeeping Division, Ministry Natural Resources and Tourism.

FPP. 2004. "Indigenous peoples' right to free prior and informed consent and the World Bank's Extractive Industries Review: An overview". http://www.forestpeoples.org/documents/law_hr/fpic_synthesis_jun07_eng.pdf

Gilligan, B. 2006. *The National Reserve System Programme 2006 Evaluation*. Canberra, Australia: Department of the Environment and Heritage. www.deh.gov.au/indigenous/publications

Githitho, A. 2003. "The sacred Mijikenda Kaya forests of coastal Kenya and biodiversity conservation". In UNESCO, 2003b, *op. cit.*

Githitho, A. 2006. "The sacred Mjiikenda Kayas of coastal Kenya: evolving management principles and guidelines". In UNESCO, 2006, *op. cit.*

Gokhale, Y. 2003. "Communicating importance of sacred groves to broader audience for conservation of biocultural heritage". Presented at the CSVPA session at the Vth IUCN World Parks Congress.

Grim, J. (Ed.) 2001. *Indigenous Traditions and Ecology: The Interbeing of Cosmology and Community.* Cambridge, MA, USA: Harvard CSWR.

Grimble, R., Chan, M.K., Aglionby, J. and Quan, J. 1995. *Trees and Trade-offs; a stakeholder approach to natural resources management.* Gatekeeper Series 52. London, UK: International Institute for Environment and Development (IIED).

Harmon, D. and Putney, A. 2003. *The Full Value of Parks: From Economics to the Intangible.* Lanham, MD, USA: Rowman and Littlefield Publishers.

Harmsworth, G. 1998. "Indigenous values and GIS: a method and a framework". *Indigenous Knowledge and Development Monitor* 6(3).

IFAD. 2005. "Free prior informed consent and beyond: the experience of IFAD". Report from an international workshop on methodologies regarding free prior and informed consent and indigenous peoples. New York, 17–19 January 2005, Department of Economic and Social Affairs, Division for Social Policy and Development, Secretariat of the Permanent Forum on Indigenous Issues.

ILO. 1989. Convention (No. 169) concerning Indigenous and Tribal Peoples in Independent Countries Adopted on 27 June 1989 by the General Conference of the International Labour Organization at its seventy-sixth session – entry into force 5 September 1991.

IUCN. 1994. *Guidelines for Protected Area Management Categories*, Gland, Switzerland and Cambridge, UK: IUCN and Cambridge University Press.

IUCN. 2008. Guidelines for Applying Protected Areas Management Categories. WCPA Draft for discussion.

Iwatsuki, K. 2006. "Sacred Forests in temples and shrines of Japan". In UNESCO, 2006, *op. cit.*

Iwatsuki, K. 2007. *Sacred sites and zoning of the Japanese Archipelago*. UNESCO, 2007, *op. cit.*

Jaireth, H. and Smyth, D. (Eds) 2003. *Innovative governance: Indigenous peoples, local communities and protected areas.* New Delhi, India: Ane Books.

Kempf, E. (Ed.) 1992. *The Law of the Mother: Protecting indigenous peoples in protected areas.* San Francisco, CA, USA: Sierra Club.

Khalid, F. and Thani, A.K. 2007. *Teacher's Guide Book for Islamic Environmental Education Promoting conservation of Misali Island – Pemba, Tanzania.* Birmingham, UK: Islamic Foundation for Ecology and Environmental Science.

Khamaganova, E. 2007. "Sacred Sites: Our pain, hope and strength". *Pachamama: a traditional knowledge newsletter of the Convention on Biological Diversity* 1(1).

Khamaganova, E. 2005. "Protection of sacred sites: global framework and local actions. Lessons from Lake Baikal and the Altai Mountains of the Russian Federation". In UNESCO, 2006, *op. cit.*

Koohafkan, P. and Boerma, D. 2006. "Conservation and sustainable management of Globally Important Ingenious Agricultural Heritage (GIAHS)". In UNESCO, 2006, *op. cit.*

Laird, S.A. 2000. *Biodiversity and Traditional Knowledge: equitable partnerships in practice.* People and Plants Conservation Series. London, UK: Earthscan.

MacDonald, K.I. 2004. "Conservation as cultural and political practice". In Borrini-Feyerabend, MacDonald and Maffi, 2004, *op. cit.*

Maffi, L. 2004. "Conservation and the 'two cultures' – bridging the gap". In Borrini-Feyerabend, MacDonald and Maffi, 2004, *op. cit.*

Malhotra, K.C., Gokhale, Y., Chatterjee, S. and Srivastava, S. 2001. *SCOPE Committee Publication.* New Delhi and Bhopal, India: Indian National Science Academy and Indira Gandhi Rashtriya Manav Sangrahalaya.

Mallarach, J. and Papayannis, T. (Eds). 2007. *Protected Areas and Spirituality. Proceedings of the First Workshop of the Delos Initiative, Montserrat, 23–26 November 2006.* Gland, Switzerland: IUCN and Montserrat, Spain: Publicaciones de l'Abadia de Montserrat.

Martin, G.J. 2003. *Ethnobotany: A Methods Manual.* People and Plants Conservation Series. London, UK: Earthscan.

Merkushina, T.P. 2007. "Role of the specially protected areas of the Khanty-Mansiysk Autonomous Okrug – Ugra in preservation of the local sacred natural-historic sites". UNESCO, 2007, *op. cit.*

Mgumia, F.H. and Oba, G. 2003. "Potential role of sacred groves in biodiversity conservation in Tanzania". *Environmental Conservation* 30(3): 259–265.

Millennium Ecosystem Assessment. 2005. *Ecosystems and Human Well-being: Synthesis*. Washington, DC, USA: Island Press.

Motonaka, M. 2006. "Sacred sites and pilgrim routes in the Kii mountain range". In UNESCO, 2006, *op. cit.*

NMK. 2004. "Sacred Mijikenda Kayas: World Heritage Nomination Dossier". Kenya: National Museums of Kenya.

Ormsby, A. 2007. "Cultural and Conservation Values of Sacred Forests in Ghana". Paper presented at the Sacred Species and Sites meeting, Cambridge Centre for Landscape and People, 23–28 September, 2007, Cambridge, UK.

Otegui, M. 2003. "The Wixarika/Huichol Sacred Natural Site in the Chihuahuan Desert of San Luis Potosi". In Harmon and Putney, 2003, *op.cit.*

Ott, M. 2000. *African theology in images*. Kachere Monograph No. 12. Blantyre, Malawi: Christian Literature Association in Malawi.

Oviedo, G. 2001."Notes on the Panel's Presentations and Discussions". *Symposium on the Importance of the Protection of Sacred Natural Sites (SNS) for the Conservation of Biodiversity, Mexico City, June 12, 2001*. Gland, Switzerland: WWF International.

Oviedo, G., Maffi, l. and Larsen, P.B. 2000. *Indigenous and traditional peoples of the world and ecoregion conservation: an integrated approach to conserving the World's biological diversity and companion map. Indigenous and traditional peoples in the global 200 ecoregions*. Gland, Switzerland, WWF International and Canada: Terralingua.

Oviedo, G. and Jeanrenaud, S. 2006. "Protecting Sacred Natural Sites of Indigenous and Traditional Peoples". In Mallarach and Papayannis, 2007, *op. cit.*

Palmer, M. and Finlay, V. 2003. *Faith in Conservation. New Approaches to Religions and the Environment*. Washington, DC, USA: The World Bank.

Papayannis, T. and Mallarach, J. (Eds). 2008. *Proceedings of the Second Delos Workshop, Ouranoupolis, Greece, October 2008*. IUCN and Med-INA, Athens, Greece.

Phillips, A. 2003. "Turning ideas on their head: the new paradigm for protected areas". *The George Wright Forum* 20(2).

Phillips, A. 2005. "Landscape as a meeting ground: Category V Protected Landscapes/Seascapes and World Heritage Cultural Landscapes". In Brown *et al.*, *op.cit.*

Pumarejo, A. and Berges, G. 2005. *Shamanism and the Forces of Nature: An analysis of the cosmovision of indigenous peoples and their sacred sites*. Proceedings of the Second International Nature Conference, April 27–30, 2005. Music for the World, WWF & IUCN.

Rabetaliana, H. and Schachenmann, P. 2003. "Past, present and future of cultural and sacred sites in Madagascar. The Importance of Sacred Natural Sites for Biodiversity Conservation". In UNESCO, 2003b, *op. cit.*

Ramakrishnan, P.S., Saxena, K.G. and Chandrashekara, U.M. (Eds). 1998. *Conserving the Sacred for Biodiversity Management*. New Delhi, India: UNESCO and Oxford and IBH Publishers.

Ramakrishnan, P.S. 2003. "Conserving the sacred: The protective impulse and the origins of modern protected areas". In Harmon and Putney, *op. cit.*

Rambaldi, G., Corbett, J., Olson, R., McCall, M., Muchemi, J., Kwaku Kyem, P., Weiner, D. and Chambers, R. (Eds). 2006. *Mapping for Change: practice, technologies and communication*. Participatory Learning and Action no 54. London, UK: IIED and Wageningen, The Netherlands: CTA. http://www.iied.org/NR/agbioliv/pla_notes/current.html

Rambaldi, G., Tuivanuavou, S., Namata, P., Vanualailai, P., Rupeni, S. and Rupeni, E. 2006. "Resource Use, Development Planning, and Safeguarding Intangible Cultural Heritage in Fiji". In Rambaldi *et al.*, *op. cit.* pp.28–35.

Ramírez, R. 1999. "Stakeholder analysis and conflict management". In Buckles, *op. cit.*

Rose, D.B. 2002. "Sacred site, ancestral clearing, and environmental ethics". In Harvey, G., 2002, *Readings in Indigenous Religions*. London, UK: Continuum.

Rössler, M. 2003. "World Heritage Sites: Toward linking the tangible and the intangible". In Harmon and Putney, *op. cit.*

Sandwith, T., Shine, C., Hamilton, L. and Sheppard, D. 2001. Reprinted in 2003. *Transboundary Protected Areas for Peace and Co-operation*. Best Practice Protected Area Guidelines Series No.7. Gland, Switzerland and Cambridge, UK: IUCN.

Secaira, E. and Molina, M.E. 2005. *Planning for the Conservation of Sacred Sites in the Context of Protected Areas: An adaptation of a Methodology and lessons from its application in the Highlands of Western Guatemala*. The Nature Conservancy.

Secretariat of the Convention on Biological Diversity. 2004. *Akwé: Kon Voluntary Guidelines for the Conduct of Cultural, Environmental and Social Impact Assessment regarding Developments Proposed to Take Place on, or which are Likely to Impact on, Sacred Sites and on Lands and Waters Traditionally Occupied or Used by Indigenous and Local Communities*. (CBD Guidelines Series). Montreal, Canada: CBD. http://www.biodiv.org/doc/publications/akwe-brochure-en.pdf

Shepherd, G. 2003. "Operationalising the ecosystem approach". Presentation to SBSTTA, Nov 2003. http://www.iucn.org/themes/cem/documents/ecosapproach/esa_g.shepherd_9thcbd_sbstta_nov2003.pdf

Shepherd, G. 2004. *The Ecosystem Approach: Five Steps to Implementation*. IUCN Ecosystem Management Series No 3. Gland, Switzerland: IUCN.

Sindiga, I. 1996. "International Tourism in Kenya and the marginalisation of the Washwhili". *Tourism Management* 17(6): 425–432.

Soutter, R., Ntiamoa-Baidu, Y., Smith, J. and Rana, D. 2003. "Recognising the contribution of Sacred Natural Sites for biodiversity conservation". Paper delivered in Workshop Stream II – Building Broader Support for Protected Areas. Vth IUCN World Parks Congress.

Stevens, S. 1997. *Conservation through Cultural Survival: Indigenous people and protected areas.* Washington, DC, USA: Island Press.

Sukhbaatar, H. 2002. *Sacred Sites in Mongolia.* ARC, WWF and World Bank.

Szabo, S. and Smyth, D. 2003. "Indigenous protected areas in Australia: incorporating indigenous owned land into Australia's national system of protected areas". In Jaireth and Smyth, *op. cit.*

Taylor and Geffin, 2003. "Battling religions in parks and forest reserves: Facing religion in conflicts over protected places". In Harmon, and Putney, *op. cit.*

Te Heuheu, T. 2006. "Culture Landscapes and the principle of guardianship". In UNESCO, 2006, *op. cit.*

Thomas, L. and Middleton, J. 2003. *Guidelines for Management and Planning of Protected Areas.* Best Practice Protected Area Guidelines Series No.10. Gland, Switzerland and Cambridge, UK: IUCN.

Tiwari, B.K., Barik, S.K. and Tripathi, R.S. 1998. "Biodiversity Value, Status, and Strategies for Conservation of Sacred Groves of Meghalaya, India". *Ecosystem Health* 4(1): 20–32.

Tucker, M.E. and Grim, J. 2001. "Series Foreword". In Grim, J. (Ed.), 2001, *op. cit.*

UNESCO. 2003a. *Convention for the safeguarding of the intangible cultural heritage.* Convention text available at http://unesdoc.unesco.org/images/0013/001325/132540e.pdf [last accessed at 20–09–2006].

UNESCO. 2003b. (Lee, C. and Schaaf, T. (Eds).) *Proceedings of an International workshop on the importance of sacred natural sites for biodiversity conservation.* Kunming and Xishuangbanna Biosphere Reserve, People's Republic of China, 17–20 February 2003. Paris, France: UNESCO.

UNESCO. 2006. (Lee, C. and Schaaf, T. (Eds).) *Proceedings of UNESCO-IUCN 'Conserving Cultural and Biological Diversity: The Role of Sacred Natural Sites and Cultural Landscapes',* Tokyo, Japan (30 May–2 June 2005).

UNESCO. 2007. (Jayakumar, R., Kim, E. and Karolyi, B. (Eds).) *The protection of sacred natural sites; importance for biodiversity conservation. Proceedings of the 10th Meeting of the UNESCO-MAB East Asian Biosphere Reserve Network, Terelj National Park Mongolia, 1–5 September 2007.*

Ver Beek, K.A. 2002. "Spirituality: a development taboo". In Eade, *op. cit.*

Verhelst, T. and Tyndale, W. 2002. "Cultures, spirituality and development". In Eade, *op. cit.*

Verschuuren, B. 2006. "An overview of cultural and spiritual values in ecosystem management and conservation strategies". Paper contributed to the International Conference on Endogenous Development and Biocultural Diversity, October 2006, Geneva, Switzerland. (Available at: http://topshare.wur.nl/naturevaluation/75146).

Verschuuren, B. 2008. "Power on this land. Sacred Sites management at Dhimurru Indigenous Protected Area in northeast Arnhem Land, Australia". Papayannis and Mallarach, 2008, *op. cit.*

Wells, M., Brandon, K. with Hannah, L. 1992. *Linking protected area management with local communities.* Washington DC, USA: World Bank, World Wildlife Fund, US Agency for International Development.

West, P.C. and Brechin, S.R. (Eds). 1991. *Resident peoples and national parks: Social dilemmas and strategies in international conservation.* Tuscon, AZ, USA: University of Arizona Press.

Wickramsinghe, A. 2005. "Adam's Peak Sacred Mountain Forest". In UNESCO, 2003b, *op. cit.*

Wickramsinghe, A. 2005. "Adam's Peak in the cultural landscape of Sri Lanka: evidence of an eco-cultural basis for conservation". In UNESCO, 2006, *op. cit.*

Wild, R.G. and J. Mutebi. 1996. *Conservation through community use of plant resources: establishing collaborative management at Bwindi Impenetrable and Mgahinga Gorilla National Parks, Uganda.* Working Paper 6, People and Plants Programme, UNESCO, Kew, WWF.

Wild, R.G., Millinga, A. and Robinson, J.M. 2007. Microfinance and environmental sustainability at selected sites in Tanzania and Kenya. Unpublished report for WWF-UK and LTS International.

Wiley, E.A. 2003. *Community-based land tenure management: Questions and answers about Tanzania's new land management act.* Issue paper no. 120. Drylands Programme. London, UK: International Institute for Environment and Development (IIED).

Xiaoxin, H. 2007. *Taking Care of Nature: Building up the Daoist Ecology Temple by our own hands: The Second Workshop on Daoism and Conservation,* July 2007, Taibaishan, China, ARC.